Guide to ATM Systems and Technology

For a complete listing of the *Artech House Telecommunications Library*,
turn to the back of this book.

Guide to ATM Systems and Technology

Mohammad A. Rahman

Artech House
Boston • London

Library of Congress Cataloging-in-Publication Data
Rahman, Mohammad A.
Guide to ATM systems and technology / Mohammad A. Rahman.
p. cm. — (Artech House telecommunications library)
Includes bibliographical references and index.
ISBN 1-58053-869-X
1. Asynchronous transfer mode. I. Title. II. Series.
TK5105.35.R34 1998
004.6'6—dc21 98-36741
CIP

British Library Cataloguing in Publication Data
Rahman, Mohammad A.
Guide to ATM systems and technology. – (Artech House telecommunications library)
1. Asynchronous transfer mode
I. Title
621.3'82'16

ISBN 1-58053-869-X

Cover design by Lynda Fishbourne

685 Canton Street
Norwood, MA 02062

International Standard Book Number: 0-89006-306-0
10 9 8 7 6 5 4 3 2 1

Contents

Foreword

Volumes of material have been written about Asynchronous Transfer Mode (ATM) since its inception. So much of what has been written pertains only to certain aspects of ATM. This is a result of the research and development that is segmented. The author believes there is a definite need for an overview that ties all these aspects together and gives a comprehensive picture of how all of the parts fit together to make up the process we call ATM. The title, *Guide to ATM Systems and Technology,* indicates the attempt to provide a broad-based approach to the entire process of ATM with emphasis on how the parts fit and work together. This approach can be beneficial to both hardware and software engineers to better understand how their contributions to ATM development meld to produce this exciting technological advance that will carry telecommunications well into the 21st century.

The purpose of this text is not to define ATM technology but to explain its different elements. If any questions arise about the real specifications of this technology, please refer to the actual standards defining the ATM technology in question.

Jerry D. Pollock

CHAPTER 1 Introduction

Introduction

The public at large will definitely think first of a cash machine when they hear "ATM." However, in the industry that makes the ATM above even possible, the term ATM has taken on a different and more exciting meaning. Asynchronous Transfer Mode (ATM) is the transport and switching technology that allows the telecommunications and computer industries to offer enhanced services to their customers. Through its fast packet and cell-based technology, ATM can provide user applications that require high-bandwidth, high-performance transport and switching. It is designed to support both constant and variable bit-rate types of user applications such as video, voice, image, and data.

Standards are being developed on the international level by the International Telecommunications Union's Telecommunications Standardization sector (ITU-T) formerly known as Consultive Committee International Telegraphy and Telephony (CCITT). American standards are being developed by the American National Standards Institute (ANSI). An association known as the ATM Forum, which represents over 250 companies throughout the telecommunications industry as well as the computing industry and some end-user representatives, is developing implementation agreements based on standards to ensure the compatibility of ATM equipment.

Broadband Switching Systems (BSS) based on ATM technology and supporting ATM interfaces will be the public carrier's chief deployment to provide a variety of services to its customers.

ATM Standards

There are two major standards bodies that are driving the development of Broadband Integrated Services Digital Network (B-ISDN) technology: the ITU-T and the ATM Forum.

ITU-T

The ITU-T is an international body that operates under the sponsorship of the United Nations and the ITU. The primary responsibility of the ITU-T is to recommend standards for the world's telephony and other public networks (PTTs) to the ITU, which sets the standards.

Narrowband Integrated Services Digital Network (N-ISDN) standards were developed by the ITU-T in the early 1980s to allow a limited capability for public networks to carry voice and data traffic. The N-ISDN standards specified two types of interfaces, namely:

1. Basic rate access at 144 Kbps;
2. Primary rate access at 1.544 Mbps or 2.048 Mbps.

These standards were not sufficient, however, for the Local Area Network (LAN) technologies (e.g., ethernet at 10 mbps).

The growing economic importance of data traffic created the widespread adoption of high-speed digital communications equipment. This indicated a need for higher bandwidth data transports. In response, study group XVIII of the ITU-T began working on B-ISDN to replace the entire public network infrastructure. Public telecommunications providers needed to upgrade their existing infrastructure of transmission, switching, and multiplexing systems and move on to an all-digital network. To accomplish this, it was necessary to define a technology that would allow all types of traffic to be carried in the one fabric.

Second, ITU-T recognized that the growth in data traffic was increasing exponentially; therefore, ITU-T had to define a technology that could far surpass the limits of then-current technologies.

After examining many different technologies, such as synchronous transfer mode (STM) and packet-switching technologies, it was decided in 1988 to base the development of B-ISDN on ATM. This was formalized by a set of initial recommendations (the blue books) published in the late 1980s. Although the term ATM is now sometimes used as a synonym for B-ISDN, it is important to know that ATM technology is one of just the possible applications that B-ISDN can use.

During the early stages of standardization, it was widely assumed that the need for ATM and B-ISDN would not affect the public network until the mid to late 1990s. It was expected that the widespread deployment of this technology would occur in the

second decade of the 21st century. This assumption proved very short-sighted, as the ATM has already entered the private network.

Although the ITU-T standards are not complete, there are sufficient standards to enable the development of early availability for commercial ATM equipment. The ANSI in America and the European Telecommunications Standards Institute (ETSI) in Europe are the two major regional bodies that generate standards based on the technical and regulatory situations of each region. These are then presented to the ITU-T.

ANSI

The ANSI committee T1 is responsible for B-ISDN on ATM standards in the United States.

ETSI

The ETSI is responsible for telecommunications standards in Europe.

ATM Forum

The ATM Forum was established in August 1991 because of the interest in ATM for private networks. The founders of the ATM Forum are as follows:

- Adaptive;
- Northern Telecom;
- Sprint;
- Cisco Systems.

Their goal was to ensure interoperability between public and private ATM implementations. The ATM Forum has grown rapidly, and today companies representing the entire networking industry have joined forces in their commitment to the widespread use of ATM technology. Some of these companies are major public and private network vendors, network providers, and major workstation vendors.

The ATM Forum adopts standards and clarifies their use. It is officially described as an "implementor's agreement" body. The main goal of the ATM Forum is to hasten availability of the ATM technology.

The ATM Forum has accelerated the adoption of ATM technology by the private networking industry. In this respect, the ATM Forum has moved quickly to set the fundamental standards necessary to allow the manufacturing of early ATM equipment for the private network. In just months after it was founded, the ATM Forum completed a specification for a set of ATM user network interfaces. Because of its ability to move more rapidly, the ATM Forum is used by U.S. public telecommunications providers (RBOCs, Bellcore and AT&T, MCI, Sprint) to define the elements of ATM services.

CHAPTER 2

Evolution of ATM

Introduction

The evolution of existing public telecommunications networks to an ATM network is based on the development and integration of digital transmission and switching technologies. ATM provides the integration of transmission and switching equipment, the integration of voice and data communication, and the integration of circuit-switching and packet-switching facilities.

It began with the revolution in data communication that prompted several new technologies for data transmission and data switching. Some of these include:

- Private lines;
- The Internet;
- Circuit switching;
- Packet switching;
- ISDN.

Circuit Switching

Circuit switching is the dominant technology today for both voice and data communications. Circuit switching implies that there is a dedicated communication path between two terminals (users). Some of the circuit-switching applications are public telephone networks, private branch exchange (PBX), and data switches within local sites. Communications by way of circuit switching require three phases:

1. Circuit establishment;
2. Data transfer;
3. Circuit disconnection.

Packet Switching

A packet-switched network is a distributed collection of packet-switched nodes. The concept of packet switching is to store information and then forward it to the next node. Messages are broken into packets. Each packet contains control information that enables the switching nodes to route packets properly. Circuit establishment is not necessary for packet switching as each packet contains routing information.

ISDN

ISDN was built on the time-division multiplexing (TDM) network to service voice, video, and data simultaneously. The primary difference between ISDN and telephony is the use of common channel signaling (out-of-band signaling) instead of in-band signaling. It is divided into two parts according to the bandwidth. These are:

1. N-ISDN;
2. B-ISDN.

N-ISDN

N-ISDN supports both basic rate interface (BRI) and primary rate interface (PRI).

BRI provides two 64-kbps bearer (B) channels and a 16-kbps data (D) channel. The B-channel is used for user data and voice transmission and the D-channel is used for control, messaging, and network management. This interface is commonly referred to as 2 B+D. This interface is intended for customer access devices, such as voice, data, and video phone.

The PRI provides twenty-three 64-kbps B-channels and a 64-kbps D-channel. The B-channel is used for user data and voice transmission and the D-channel is used for control, messaging, and network management. This interface is commonly referred to as 23 B+D in North America. Internationally, it is 30 B+D. This interface is intended for higher bandwidth access, such as PBX and LAN. In North America, twenty-four DS-0 channels of 64-kbps each constitute a DS-1 stream. Therefore, a single DS-1 at the rate of 1.544 Mbps is also used for PRI.

B-ISDN

Rapid growth in data traffic, the need for upgrading existing equipment, and the move to all-digital networks accelerated B-ISDN development. Before ATM was finalized by a set of recommendations published in late 1980s, some of the following technologies were examined:

- Packet switching;
- STM;
- ATM.

ATM is just one of the possible applications that B-ISDN can use.

Digital Data Transmission Method

In a digital communications system, there are two types of data transmission methods used for data transmission:

1. Asynchronous data transmission;
2. Synchronous data transmission.

Asynchronous Data Transmission

The asynchronous data transmission method does not use a clock with the transmitted datastream. As seen in an ASCII character transmission, each character is encapsulated by start and stop bits. Figure 2.1 shows an 8-bit character transmission with a start and a stop bit. Time between each character transmission is not specified, and they do not have to be equal.

Synchronous Data Transmission

A clock is essential to capture the bits at a constant rate for synchronous data transmission. The clock is either associated with, or driven from, thc digital datastream. Both the transmitter and receiver use the same clock signal or derive the clock signal within a certain frequency tolerance. Normally, parallel interfaces use a separate clock line. Figure 2.2 shows a typical synchronous data transfer where the message begins with a synchronization (SYNC), start of message (SOM), and control characters and ends with cyclic redundancy check (CRC) and end-of-message (EOM) characters. Synchronous data transmission usually operates at higher speeds than asynchronous data transmission.

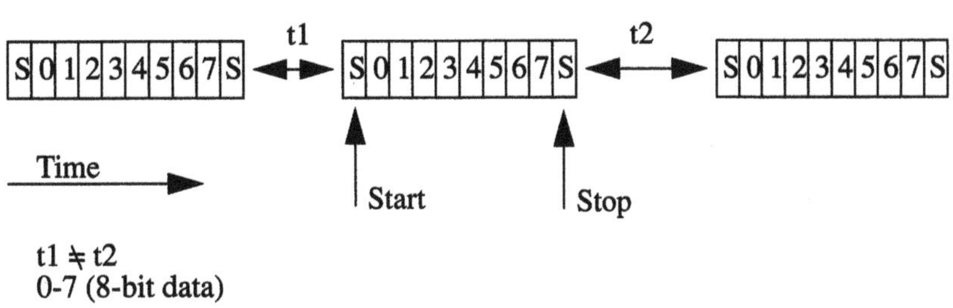

FIGURE 2.1. Asynchronous character transmission.

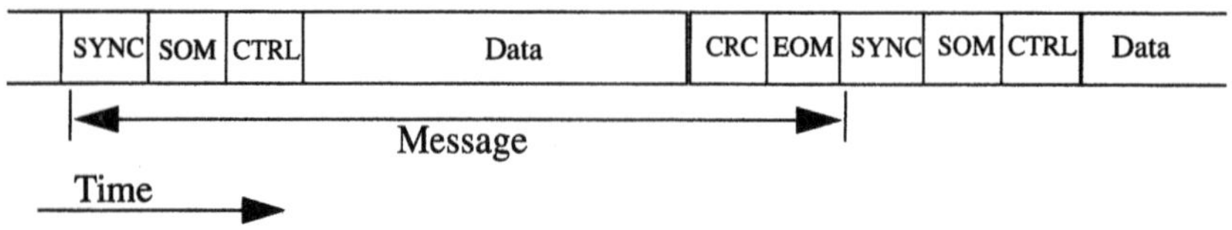

FIGURE 2.2. Synchronous data transmission.

Digital Data Transmission Mode

There are two data transmission methods that are generally used in the synchronous data transmission environment:

1. ATM;
2. Synchronous transfer mode (STM).

ATM

The ATM is also known as *asynchronous time-division multiplexing*. ATM is more suitable for a packet-switching environment. A header field always prefixes each fixed-length channel (payload), as shown in Figure 2.3. The header field contains all the information relating to the channel attached to it.

Therefore, these timeslots can be used by any user who has data ready to transmit. Unassigned (idle) cells (packets/cells) occupy timeslots if no users are ready to transmit. Therefore, a user can utilize multiple timeslots to transmit data

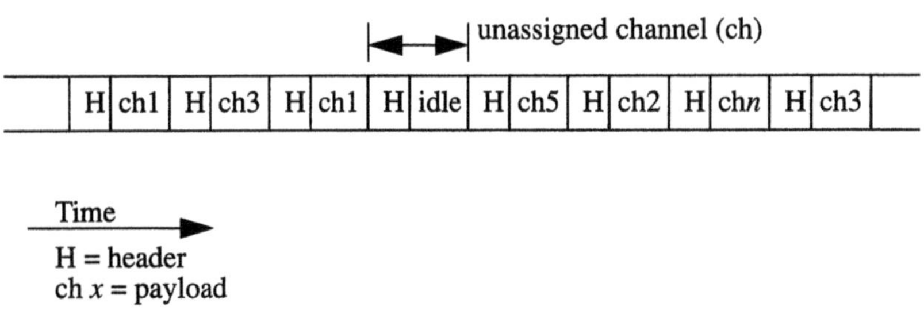

FIGURE 2.3. ATM.

when needed. ATM is much more efficient for the variable traffic load when compared with STM.

STM

The STM is also known as *synchronous time-division multiplexing*. Unlike ATM, STM channels are bound by a frame, and each timeslot within a frame is reserved to a single channel as shown in Figure 2.4. These channels repeat in each frame in their respective timeslots. Thus, if a channel is not transmitting data, the timeslot is wasted, and in case other channels have more data to transmit, they have to wait until their dedicated timeslot occurs again.

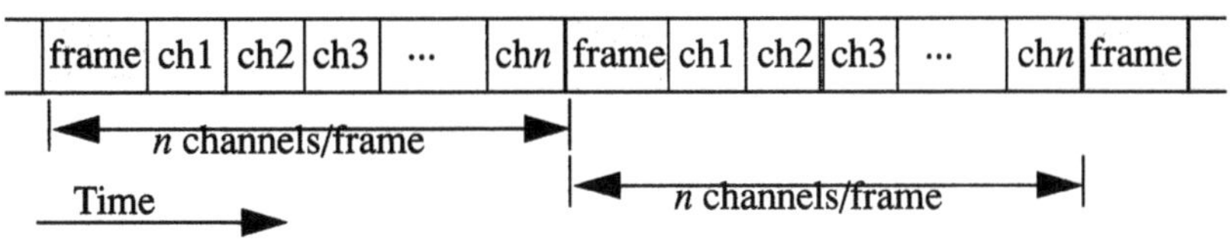

FIGURE 2.4. STM.

CHAPTER 3

ATM Network Architecture

ATM Network

The ATM standards define two major types of interfaces in ATM networks. These are as follows:

1. User-to-network interface (UNI);
2. Network-to-network interface, or network-to-node interface (NNI).

An ATM network is composed of end users; that is an ATM card on a router and intermediate switches, all linked by transmission media. Figure 3.1 shows the ATM network structure. The hierarchical organization used in existing telephone systems is not used in the ATM network. The ATM standards define the protocols used across the links connecting the terminals and the switches. The UNI was intended to be the interface between an ATM terminal device and a public ATM switch because the ATM standards were originally defined for the public network. A comparison may be drawn with the interface between a telephone and the central office. The UNI now represents a regulatory boundary between a public network and customer premises equipment (CPE).

The NNI is the interface between public switches. Figure 3.2 shows the public and private ATM switches in the ATM network. The term UNI refers now to any interface between an end user and an ATM switch (public or private).

The term NNI, generically, is used to identify a switch-to-switch interface but does not include the public-to-private ATM switch UNI. Interswitching system interfaces (ISSI) are referred to as switch-to-switch links. The ISSI can be categorized as follows:

- Private ISSI (for links between private ATM switches);
- Public ISSI (for links between different public switches).

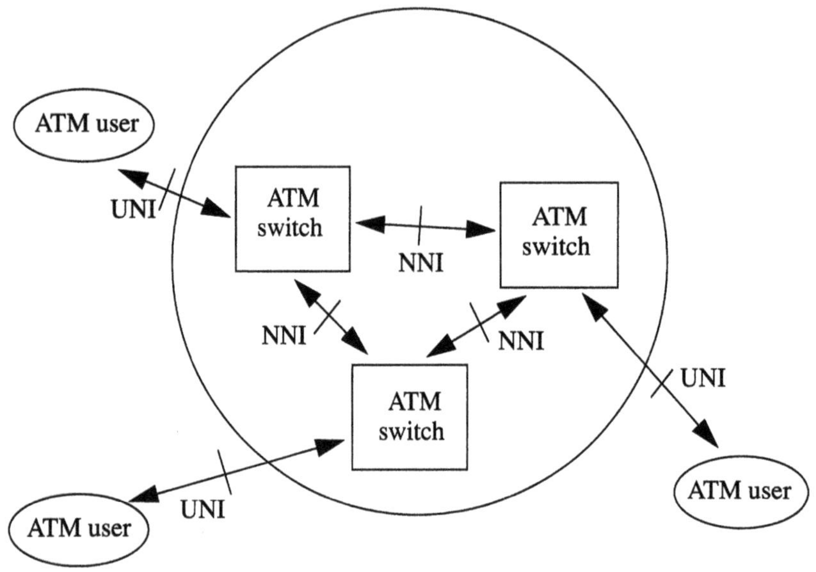

FIGURE 3.1. ATM network.

Public ISSI can be further categorized as follows:

- Intra-LATA (local access and transport area) ISSI, where both public switches and the links between them belong to the same service provider;
- Inter-LATA ISSI, or inter-carrier interfaces (ICI), which are used to link the ATM switches of a service provider with those of an interexchange carrier (e.g., AT&T, Sprint, MCI).

The UNI carries aggregate user traffic from existing applications, such as PBX tie trunks, host-to-host private lines, and videoconference circuits. It also interconnects multimedia high-speed networks, such as workstations, supercomputers, routers, and bridges. There are two types of ATM UNI defined by the ATM Forum:

1. Public UNI;
2. Private UNI.

The public UNI is typically used to interconnect ATM users (e.g., a router) with an ATM switch located in a public service provider's network. On the other hand, the private UNI is typically used to interconnect an ATM user with an ATM switch located in a private ATM network (e.g., an ATM switch managed by private corporations or universities).

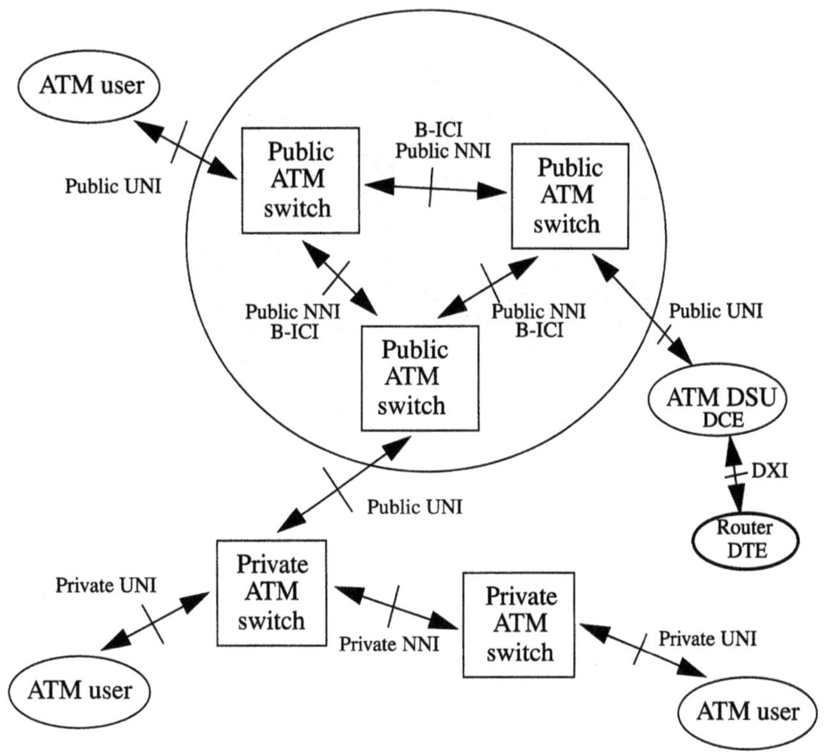

FIGURE 3.2. ATM network (public and private ATM switch).

Both UNIs share the same ATM layer specification, but physical media may be different. Because of different application requirements, the functionality may differ between the public and private UNI. The private ATM switching equipment may collocate with user devices; therefore, the distance between user device and the switch may be very short and hence can use limited distance technology. On the other hand, user devices connected with the public ATM switch must be capable of spanning long distances. Figure 3.3 shows the locations of the ATM UNI. The UNI allows the transfer of user application information.

The ATM data exchange interface (DXI) was developed to allow current routers to interwork with ATM networks without requiring special hardware. The data terminal equipment (DTE) (e.g., router) and the data communications equipment (DCE) (e.g., ATM data service unit (ATM DSU)) together provide a UNI.

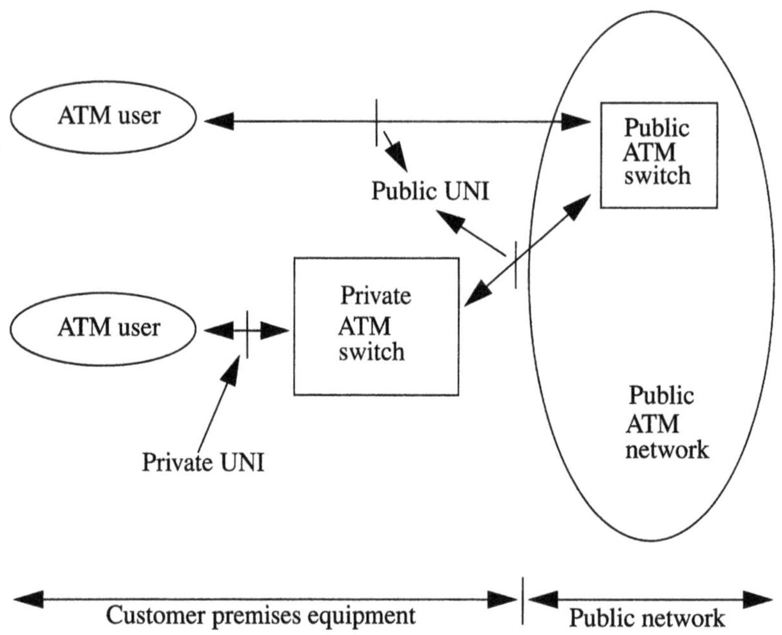

FIGURE 3.3. Locations of the ATM UNI.

There are two types of ATM NNI defined by the ATM Forum:

1. Public NNI;
2. Private NNI.

The public NNI, also known as the broadband intercarrier interface (B-ICI), is used to interconnect public ATM switches. On the other hand, the private NNI (P-NNI) is used to interconnect ATM switches, which are managed by a private network.

Refer to ATM Forum documents on ATM UNI specifications, ATM DXI specifications, and ATM B-ICI specifications for more information.

ATM Protocol

The ITU-T has created a layered model of the ATM protocol defined in ITU-T recommendation I.321. Figure 3.4 shows the ATM protocol reference model. The bottom layer is the physical layer. This layer is divided into two sublayers, namely, the transmission convergence sublayer and the physical media-dependent sublayer. The ATM layer and the ATM adaptation sublayer are situated above the physical layer.

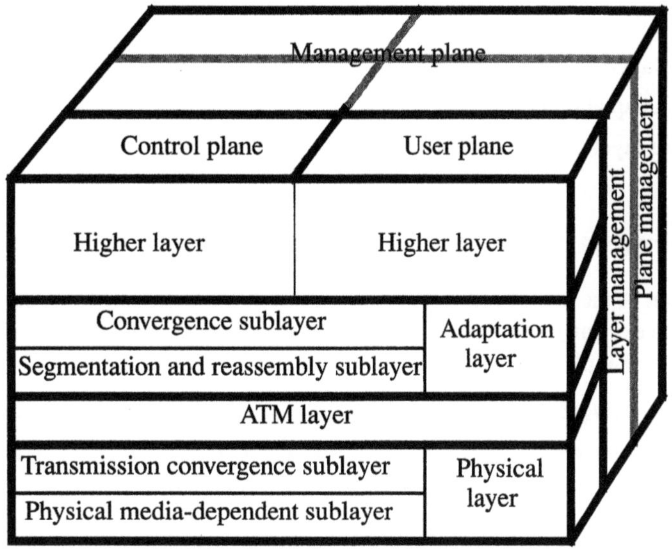

FIGURE 3.4. ATM protocol reference model.

These two layers are equivalent to the open systems interconnection (OSI) data link layer. An OSI is a seven-layer model defining the international protocol standards for data communications in a multiple architecture and vendor environment.

The ATM protocol model is divided into three planes as described below.

User Plane

The user plane facilitates end-to-end or user-to-user data transfer. The user plane services the following layers:

- Physical layer;
- ATM layer;
- ATM adaptation layers.

Control Plane

The control plane supports signaling across ATM layers for call establishment, release, and other connection control functions necessary for providing switched services. The control plane shares the following layers with the user plane:

- Physical layer;
- ATM layer;
- ATM adaptation layers;
- Higher layer signaling protocols.

Management Plane

The function of the management plane is not only to manage functions, but also to facilitate the exchange of information between the user plane and the control plane. The management plane is further subdivided into two sections:

1. *Layer management*, which performs layer-specific management functions;
2. *Plane management*, which performs management and coordination functions related to the complete system.

The functions performed by each ATM protocol layers are summarized below.

The ATM layer relays cells in a switch or recognizes cells as belonging to a known circuit and passes them on to the end user. The function of the physical layer is to pass a stream of cells to the ATM layer; similarly, the ATM layer passes a stream of cells back to the physical layer. The ATM adaptation layer performs a convergence function between applications using the ATM network and the ATM layer; that is, the AAL layer generates ATM cell payloads from the data passed to it by the higher layers using the ATM network. Figure 3.5 shows the ATM protocol layer functions.

The descriptions in the following chapters are based on the most current ITU-T recommendations, ANSI standards, and the latest specifications of the ATM Forum UNI standard and the Internet Engineering Task Force (IETF) RFCs related to ATM. Since standardization efforts are ongoing, it is likely that some of what is described will change, and the latest drafts of the relevant standards should be consulted for the most accurate information.

The ATM Forum UNI specification specifies the lower two layers of the B-ISDN protocol reference model. These two layers (the physical layer and the ATM layer) deal with the transport of information between two users. Since the ATM Forum is the driving force for the development of the ATM network, we choose to follow the standards of the ATM Forum for the following two chapters. These two chapters contain detailed information on the physical layer and the ATM layer.

Higher layer	Higher layer
Convergence sublayer	Adaptation layer
Segmentation and reassembly sublayer	
ATM layer Generic flow control Cell header generation and extraction Cell VPI/VCI translation and compression Cell multiplex and demultiplex	ATM layer
Transmission convergence sublayer Cell rate decoupling HEC sequence generation and verification Cell delineation Transmission frame adaptation Transmission frame generation and recovery	Physical layer
Physical media-dependent sublayer Bit timing Physical medium	

FIGURE 3.5. ATM protocol layer functions.

CHAPTER 4

Plesiochronous Digital Hierarchies

Digital Time-Division Multiplexing

Plesiochronous digital transmission is widely used by public networks for voice applications. It is also used for data communications. Recently the synchronous optical network (SONET) in North America and the synchronous digital hierarchy (SDH) in other countries developed to support high-speed and high-quality digital transmission.

Refer to Bellcore GR-499-CORE for more information.

Plesiochronous Digital Hierarchies

Plesiochronous means *nearly synchronous*. It was developed by the Bell Labs to carry digitized voice efficiently and evolved into the North American digital hierarchy, as shown in Table 4.1. The lowest level is known as DS-0 at a rate of 64 kbps. In the hierarchy, the lower numbered digital streams are multiplexed into the higher numbered digital streams within a certain frequency tolerance.

Twenty-four DS-0 streams are combined to create a DS-1 stream. Four DS-1 streams are multiplexed together to create a DS-2 stream, and seven DS-2 streams are multiplexed to create a DS-3 stream. The bit rates are not the exact multiple of the lower level of DS-*n* because of the extra overhead information required for each level of the digital hierarchy.

TABLE 4.1. Plesiochronous Digital Hierarchies

Signal name	Bit rate	Number of DS-0s
DS-0	64 kbps	1
DS-1	1.544 Mbps	24
DS-2	6.312 Mbps	96
DS-3	44.736 Mbps	672

DS-0 Signal

The lowest level of the plesiochronous digital hierarchy is known as DS-0 at a rate of 64 Kbps. DS-0 signals are always embedded within a DS-1 bitstream.

Synchronous Digital Data

Synchronous digital data provide full-duplex, synchronous, end-to-end digital transmission for the signal rates of 2.4, 4.8, 9.6, 19.2, 56, and 64 Kbps on dedicated two-point and multipoint circuits. Unstructured serial data from the customer at 2.4, 4.8, 9.6, 19.2, and 56 Kbps is structured into DS-0 octets, and customer data at the 64-Kbps rate is octet-structured by the customer.

Voice

The telecommunications digital transport systems predominantly use a form of pulse modulation known as *pulse code modulation (PCM).* The following paragraphs describe how a DS-0 signal is created from analog voice frequency.

The natural voice frequency ranges from 300 to 3,300 Hz, where most of the information in a voice is conveyed by frequencies around 1,000 Hz. In the digital networks, analog voice signals are converted to digital bitstreams and treated like channel data (payload). Analog signals can take on any value within a wide limit. Digital signals are limited, however, in taking on only discrete values because of the bandwidth limitation (number of bits that can be transmitted economically).

Figure 4.1 shows the analog-to-digital conversion process. The analog voice signal is shown as the solid curve, which takes on any value within a wide range. Digital signals are the sampled point on the curve at a constant rate. The binary values are the closest digital representation of the points on the curve because the digital signals are limited to a few steps. The difference between the actual value and the closest digital representation is known as the *quantizing noise.*

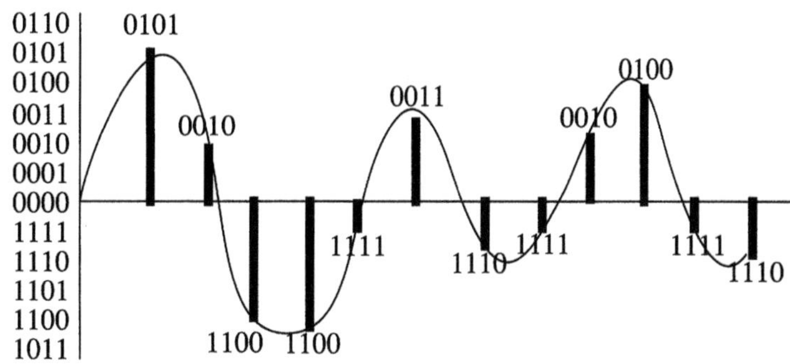

FIGURE 4.1. Analog-to-digital conversion.

There are several ways to digitize voice. Because the world standard for digital voice is PCM, it will be discussed in this chapter.

PCM

The analog voice signal is sampled 8,000 times per second. According to the Nyquist theorem, the sampling rate must be two times the highest frequency of the original signal to avoid *aliasing* (representation of more than one analog signal from the resulting digitized points). The modulator uses the sample to send a very narrow, square wave pulse representing the same analog signal at that point. This process is called *pulse-amplitude modulation (PAM).*

The height of the pulse is then converted to a digital value by a coder (analog-to-digital converter). The 8-bit code word (pulse code) representing the voltage of the pulse at the sample point is created. This process is known as *digital encoding.*

The two-step process described above creates a digital stream of 64 kbps (8 bits x 8,000 samples/sec) from an analog voice signal.

Nonlinear voice encoders were introduced to save digital bandwidth and better representation of analog voice signal by using only 8 bits (256 different values). This nonlinear ruler (voice ruler) has finer spacing at low volume levels and wider spacing at louder volume levels as shown in Figure 4.2. The original voice signal is compressed for transmission, then expanded at the receiving end. This two-step process is called *companding.*

In North America and Japan, this companding is known as "mu-law"; it is "A-law" in the rest of the world. The nonlinear voice encoders are different for A-law PCM and mu-law PCM. Refer to ITU-T recommendation G.711 for PCM (A-law, mu-law) of voice requirements.

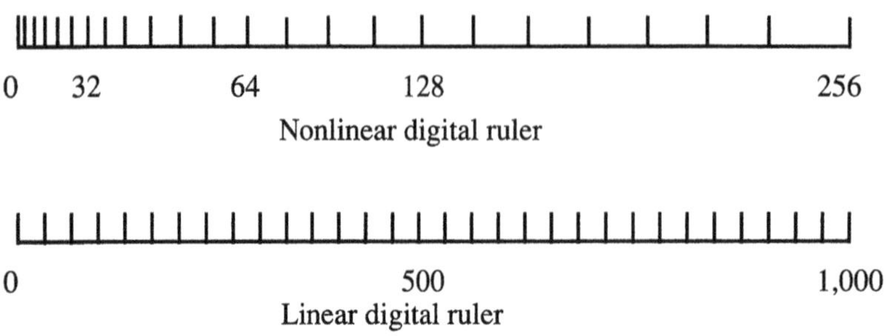

FIGURE 4.2. Nonlinear versus linear ruler.

At the receiving end, the 64-kbps digital datastream is divided into 8-bit words, which are then decoded to analog values. Narrow analog output pulses are created every 1/8,000 sec using the nonlinear voice ruler. These recovered pulses are the same as the pulses created by the transmitter PAM modulator. A lowpass filter then smooths the output into a continuous varying analog signal, which is very similar to the original voice signal at the transmitting end.

Voice Channel Bank

The voice channels are digitized and multiplexed in PCM voice channel banks. Twenty-four individual voice signals are passed through 24 individual PAM blocks. The pulse streams from the 24 channels are multiplexed (interleaved) on a single analog PAM bus. Then, the digital coding is done by one codec, between the analog PAM bus and a high-speed digital (T1) line.

When receiving, the channel bank's codec converts T1 bitstream into pulses interleaved on the PAM bus. Individual PAM modules capture their own pulses during their own timeslot and re-create the original voice signal.

Telephone Signaling

The "off hook" (when a caller picks up the telephone) request for service must be carried to the central office (i.e., telephone switch). Likewise, "on hook," dial pulses, and busy indication are called *telephone signaling*. Signaling over the T1 span is done in bits. The presence of a specific signaling condition is coded and multiplexed with the voice information.

A small portion of the digital voice (i.e., PCM) is used to carry signaling information with no apparent effect on voice quality. The least significant bit (LSB) in every sixth PCM sample in each voice channel is reserved for signaling. This is

also known as robbed-bit signaling. These bit positions are not available for voice information.

PCM Channel Rate

The PCM channel rate is 64,000 bits/sec, which includes voice and telephone signaling information. Each PCM channel contains 8 bits. Therefore, a single bit in a PCM channel repeats 64,000/8 = 8,000 times per second.

Since the LSB of every sixth PCM sample in each voice channel carries signaling information, the PCM channel includes 8,000/6 = 1,333 bit/sec for robbed-bit signaling (the LSB in every sixth PCM sample in each voice channel is replaced with a signaling bit). This implies a pure voice rate at 64,000–1,333 = 62,667 bit/sec.

To avoid dealing with the signaling bit, the LSB of each PCM sample is not used for data transmission. Therefore, the data rate = 64,000–8,000 = 56,000 bits/sec.

DS-1 Signal

Twenty-four DS-0 streams are combined to create a DS-1 stream. The nominal rate of DS-1 signal is 1.544 Mbps.

D4 Framing

The standard D4 format (also called M24 multiplexer format) contains 24 timeslots (channels). Each timeslot is represented by an 8-bit digital word (PCM). The frames repeat at the sampling rate of 8,000 times per second or once every 125 μs. Thus, one channel per frame represents 8 x 8,000 = 64,000 bit/sec. These channels are also known as DS-0 (digital signal level zero) worldwide, or the standard voice channel.

The data in one frame is 8 bits x 24 channels (timeslots) = 192 bits long. In addition to 24 channels, there is one framing bit used per frame that marks the start of the framing sequence. Thus, a channel bank transmits a series of bits at a rate of 1.544 Mbps on the T1 line. This is also known as DS-1 (digital signal level one). The D4 frame format is shown in Figure 4.3.

Outside North America and Japan, 32 DS-0s constitute an E-1 (DS-1) of 2.048 Mbps. Because one of the DS-0 channels is reserved for synchronization, there is no additional framing bit required. A second DS-0 carries only signaling information. Therefore, the number of voice channels is reduced to 30 per E-1 (DS-1).

Because the framing bit marks the beginning of the 24-timeslot sequence, it must be recognized in a high-speed bitstream. To locate the framing bit, the receiver looks for a fixed sequence that repeats every 12 or 24 frames. There are two types of framing standard supported that were first introduced by AT&T:

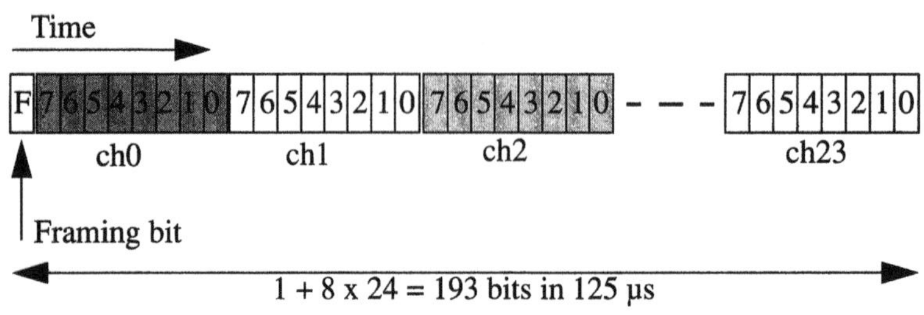

FIGURE 4.3. D4 frame format (single frame).

1. Superframe;
2. Extended Superframe.

Superframe

Twelve D4 frames constitute a superframe (SF) as shown in Figure 4.4.

A framing bit separates each frame from the other. The fixed sequence of the framing bits (100011011100) identifies the start of the superframe. The LSBs of all 24 voice channels are used for robbed-bit signaling in the sixth and twelfth frames of the SF as shown in Figure 4.5. Signaling bits in the sixth frame and the twelfth frame are called "A" and "B" bits, respectively. Only four signaling conditions can be conveyed using these bits. In most cases, these two bits are set the same. The sequence of transmission between them can control or indicate disconnect, busy, and dial pulses. The two supervisory states are:

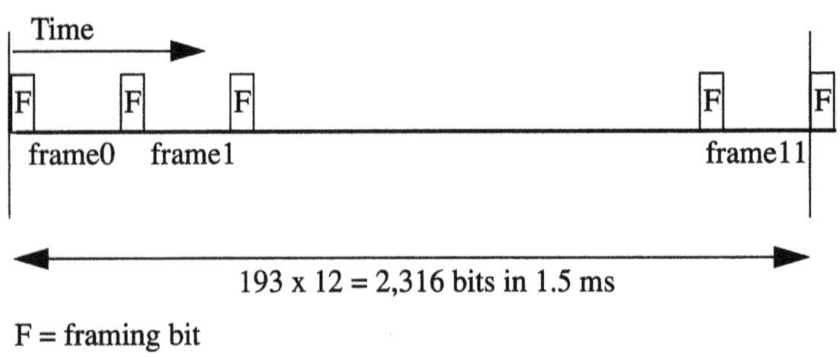

FIGURE 4.4. Single superframe (SF).

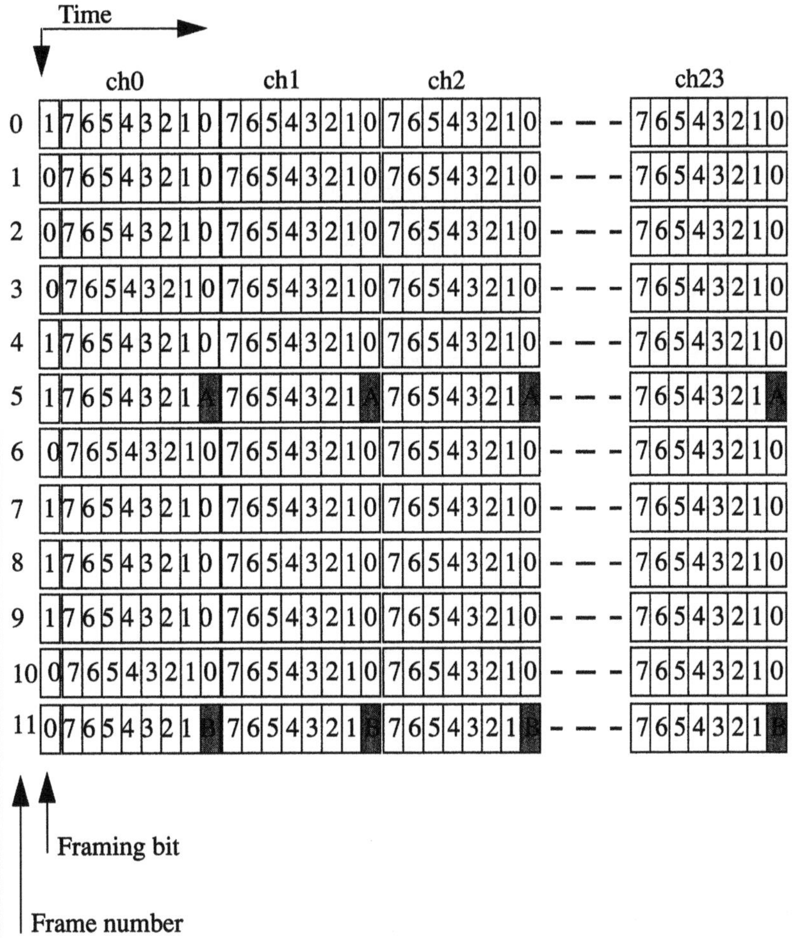

FIGURE 4.5. Robbed bit signaling (SF mode).

1. On-hook: A = 0, B = 0;
2. Off-hook: A = 1, B = 1.

As an example, Figure 4.6 shows a DS-1 signal is being created from a channel memory where PCM signals are stored, a supervisory memory where "on-hook" and "off-hook" signals are stored, and a frame-pulse sequence (FPS) register where the SF framing sequence is stored. To create a DS-1 signal in SF format, the first framing bit from the FPS register is taken followed by 24 PCM samples beginning with the ch0 from the channel memory. Then the second framing bit from the FPS register is taken, followed by 24 PCM samples beginning with the ch0 from the channel memory. PCM samples located in the channel memory are updated every 125 μs. This process is repeated for the twelve D4 frames to create a D4 superframe

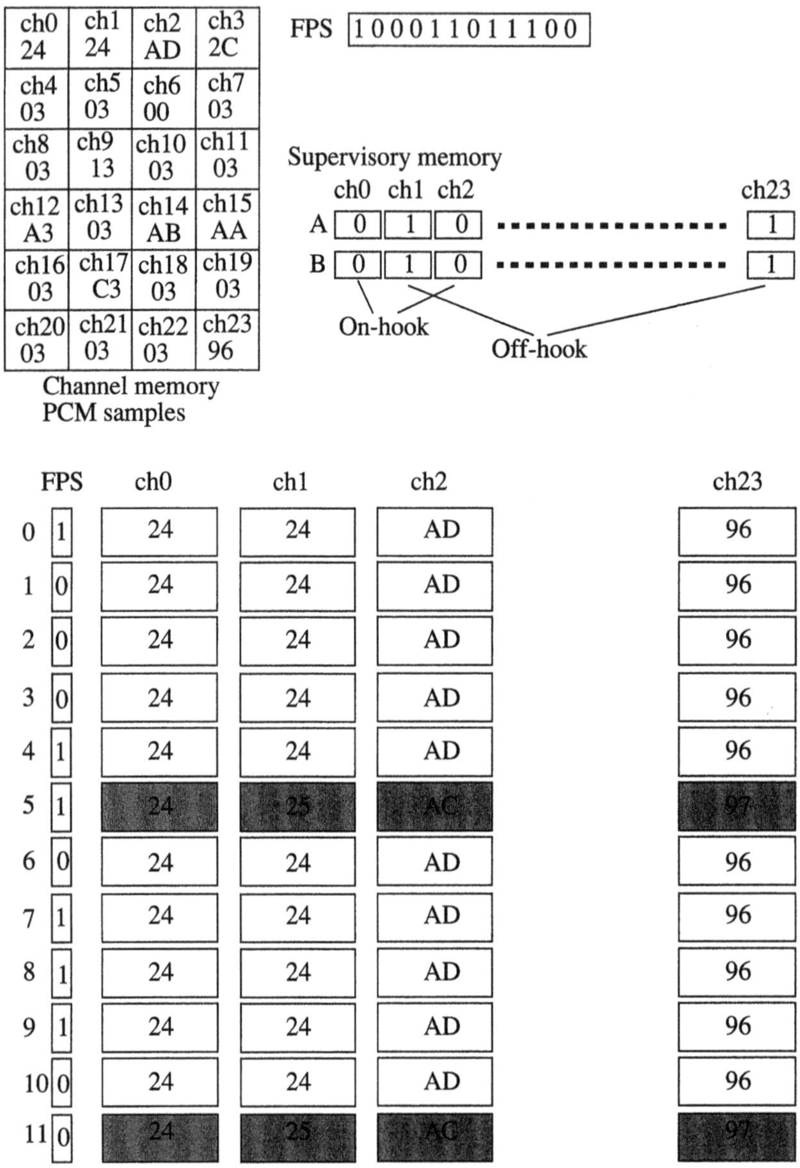

FIGURE 4.6. DS-1 signal generation (SF mode).

with the exception of the sixth and twelfth frame, where LSBs of all 24 channels of the D4 frame is replaced by the A (sixth frame) and B (twelfth frame) bits from the supervisory memory, which are constantly updated to reflect the current status of each of the DS-0 channels within the DS-1 signal.

Extended Superframe

The extended superframe (ESF) was introduced by AT&T to accommodate 16-state (A, B, C, D) signaling, CRC-error checking (6-bit CRC), and DL-diagnostic data channel (12 bits) beside FPS-synchronization (6 bit). Twenty-four D4 frames are used to create an ESF that doubles the length of the older 12-frame SF. The F-bit sequence is 24 bits long, which still identifies every sixth frame for robbed-bit signaling. The ESF is shown in Figure 4.7. The robbed-bit signaling for ESF format is shown in Figure 4.8.

The number of D4 frames is doubled to make room for two additional signaling bits, C and D, in frames 18 and 24. The A- and B-bits are repeated in C- and D-bit positions in most cases.

A 2-kbps framing pattern sequence (FPS) -001011 is used to identify the frame and the ESF boundaries. A 2-kbps CRC sequence carries the CRC-6 code. Beginning with the F-bit of the first frame, every other F-bit of an ESF is used to form a 4-kbps facility data link (FDL) channel.

Two separate formats are used for message transmission on the FDL channel. The first format is the scheduled message that uses a 15-octet HDLC packet to send facility performance information. The second format is a group of unscheduled messages that use repeated 16-bit codewords to send alarms, commands, and responses.

Figure 4.9 shows that a DS-1 signal is being created by multiplexing 24 PCM data, which are sourced from 24 different voice phones. A framing bit is added at the beginning of each set of twenty-four 8-bit PCM samples. The frame pattern is stored in a register. The signaling bits replace the least significant bits of each PCM

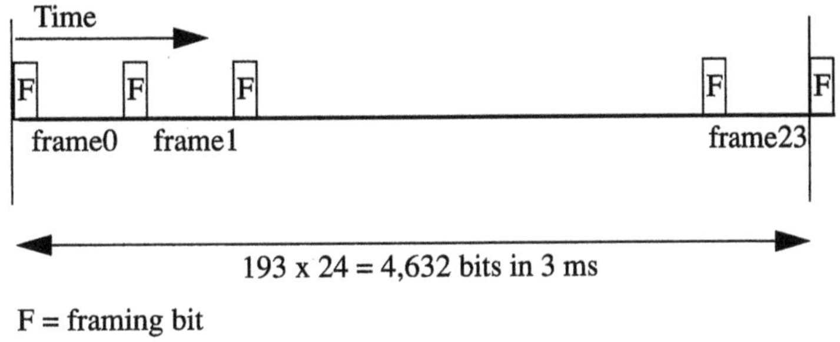

FIGURE 4.7. Single ESF.

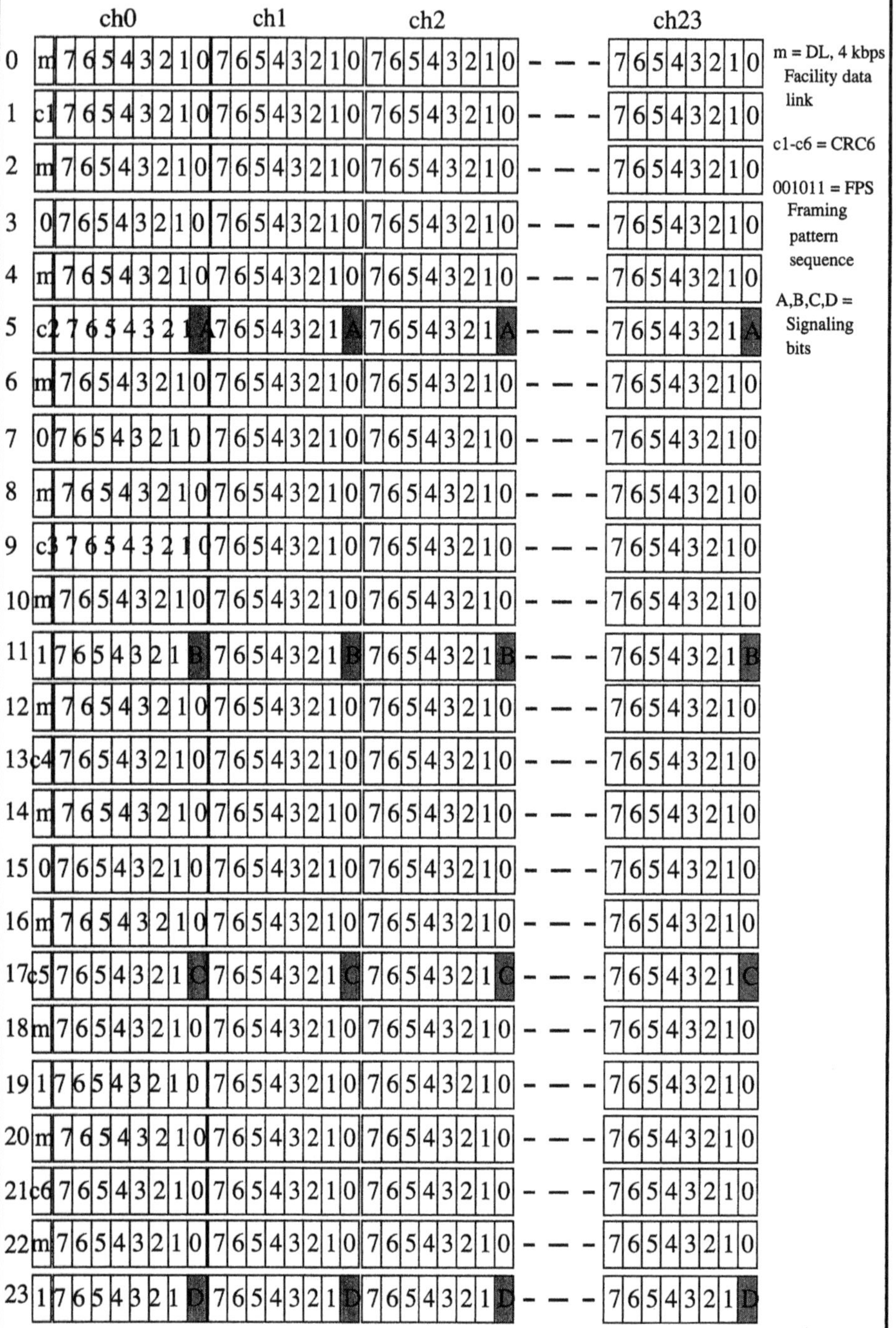

FIGURE 4.8. Robbed-bit signaling (ESF mode).

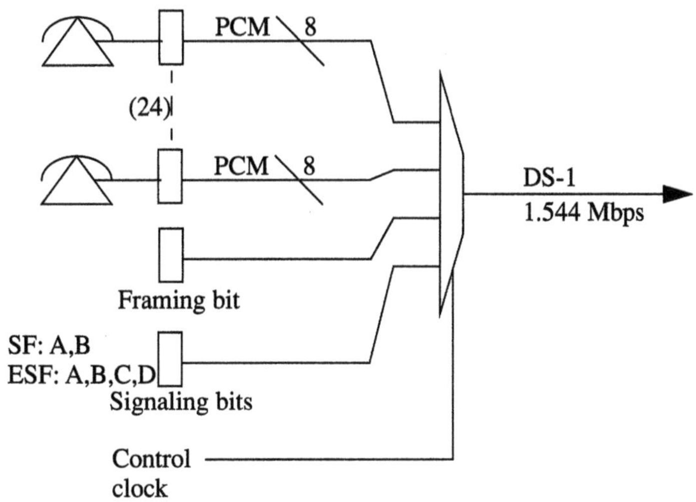

FIGURE 4.9. DS-1 signal generation.

sample of the signaling frame that is every sixth frame. The signaling information is stored in the memory, and it represents the status of each telephone.

User Data Framing

The framing of the user data depends on the networking situation and the user's applications. To meet the minimal requirement, every 193rd bit position must follow an SF or ESF pattern. This is true regardless of the information format sent by the user. The information may be ATM cells at the rate of 1.536 Mbps (192 bits/125 μs). ATM cells are 53 octets long, longer than a D4 frame (24 octets plus a framing bit). Therefore, it takes more than two D4 frames to transport a single ATM cell.

D4 framing (F-bits in the 193rd position) may be inserted in the synchronization character on a proprietary frame-based transmission or in the ATM packet. D4 framing is transparent to this type of multiplexer function because the insertion of F-bits at the transmitting end, and their removal at the receiving end, is done in the hardware. ATM over DS-1 is shown in Figure 4.10.

DS-2 Signal

The nominal DS-2 interface rate is 6.312 Mbps. Traditionally, four DS-1 signals are multiplexed into a single DS-2 signal.

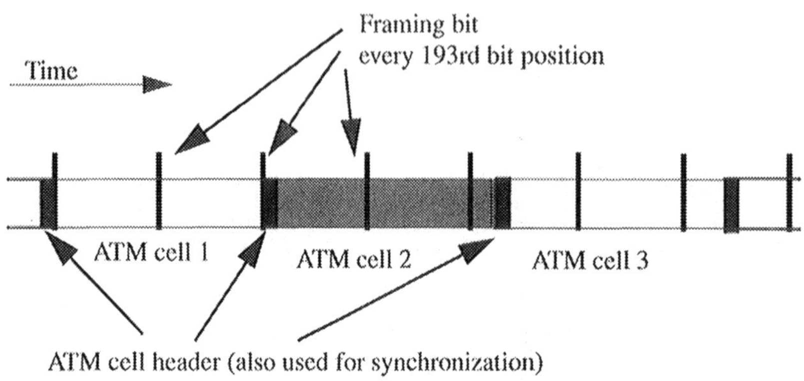

FIGURE 4.10. ATM over DS-1.

DS-2 M-Frame Format

The DS-2 signal is divided into M-frames. Each M-frame contains 1176 bits. The M-frames are divided into four M-subframes consisting of 294 bits. Each M-subframe is further subdivided into six blocks consisting of 49 bits. The DS-2 M-frame structure is shown in Figure 4.11.

In the M12 multiplexing mode, stuffing for first DS-1 channel occurs in first M-subframe, in the first information bit of the last block. Stuffing for second DS-1 channel occurs in second M-subframe, in the second information bit of the last block. Stuffing for third DS-1 channel occurs in third M-subframe, in the third information bit of the last block. Stuffing for fourth DS-1 channel occurs in fourth M-subframe, in the fourth information bit of the last block. All three C-bits are set to 1 if stuffing occurs in that M-subframe. Indication of no stuffing is conveyed by setting all three C-bits to 0.

DS-3 Signal

Traditionally, twenty-eight DS-1 signals are multiplexed into a single 44.736-Mbps DS-3 signal (M13 multiplex). Today, telecom networks have a significant investment in DS-3 transmission equipment. Therefore, ATM mapping for DS-3 is very important.

M-Frame Format

The DS-3 signal is partitioned into M-frames of 4,760 bits each. The M-frames are divided into seven M-subframes, each having 680 bits. Each M-subframe is then

TABLE 4.2. C-bit Functions and Values. (*After:* GR-499-CORE, Bellcore.)

M-subframe number	C-bit number	Function	Value
1	C1 C2 C3	Application identification Reserved FEAC	1 1 FEAC
2	C4 C5 C6	Unused	1 1 1
3	C7 C8 C9	CP (parity)	P P P
4	C10 C11 C12	C10 Far end bloc error (FEBE)	FEBE FEBE FEBE
5	C13 C14 C15	Terminal/terminal data link	DL DL DL
6	C16 C17 C18	Unused	1 1 1
7	C19 C20 C21	Unused	1 1 1

(FEAC) signals are encoded into repeating 16-bit codewords for transmission. Codewords look like the following:

- 0 X X X X X X 0 1 1 1 1 1 1 1 1

The right-most bit of the codewords is transmitted first. The six bits labeled *x* can represent 64 distinct codes (signals). Bellcore document GR-499-CORE defines items, such as assigned alarm, status, and loopback control codeword.

The main information timeslots are organized as 8 rows and 84 columns instead of the linear representation of a series of 7 rows and 85 columns of M-frames. Auxiliary timeslots are added to carry DS-1 framing bits. The alternative view of the M-frame is shown in Figure 4.15.

Figure 4.16 shows the DS-3 signal payload decomposition.

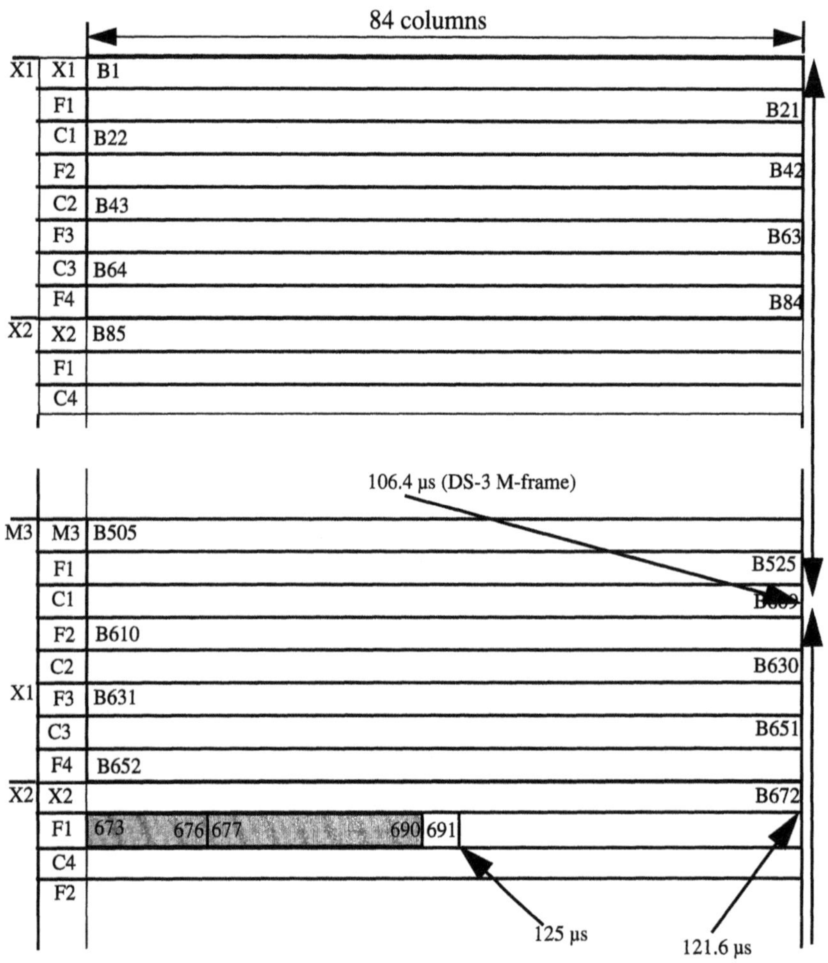

X1 B1-B84
X2 B85-B168
P1 B169-B252
P2 B253-B336
M1 B337-B420
M2 B421-B504
M3 B505-B588
X1 B589-B672
X2 B673-B691, B1-B65

(9 overhead + 690 info) x 8 = 5,592 bits

5,592 bits/44.736 Mbps = 125 µs

B = eight bits.

B673 through B676 carry DS-1 framing bits in the byte-synchronous case.

B673 through B676 carry DS-1 info bits in the bit-asynchronous case.

B677 through B690 carry DS-1 signaling bits in the byte-synchronous case.

B677 through B690 carry DS-1 stuff control and stuff bits in the bit-asynchronous case.

B691 unspecified octet; this is used to match 125 µs boundaries in syntran mode.

Note: 699 x 106.402 µs = 595 x 125 µs

FIGURE 4.15. DS-3 M-frame (alternative view).

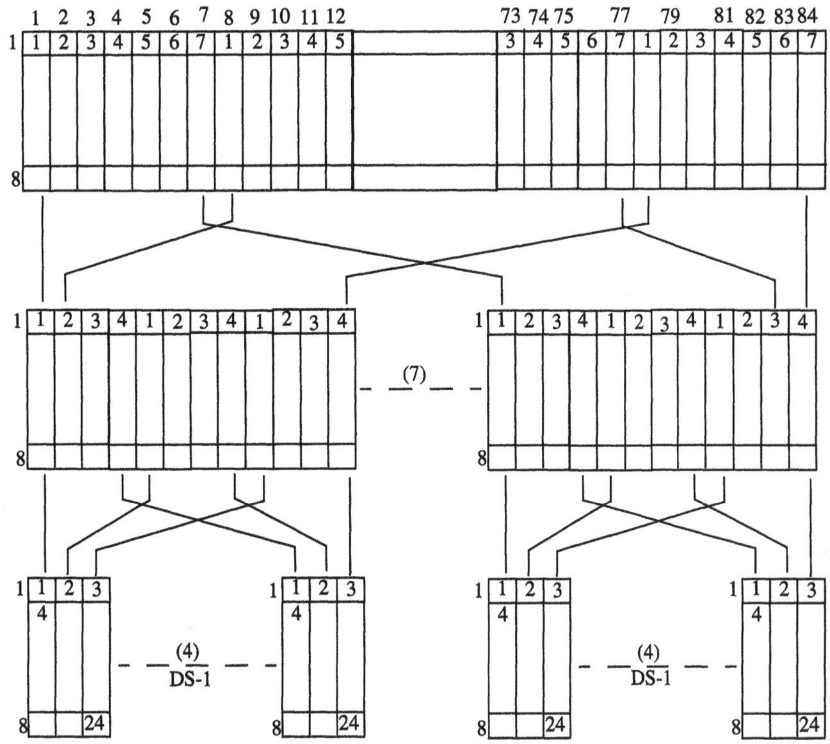

24 DS-0 x 28 DS-1 = 672 DS-0
Note: Each position represents a bit from a DS-0 sample.

FIGURE 4.16. DS-3 signal payload decomposition (M23 mode).

CHAPTER 5

SONET

Introduction

SONET in North America and the Synchronous Digital Hierarchy (SDH) by the ITU-T are primary choices of a physical medium for ATM. In the United States, the SONET standard, ANSI T1.105, was released in 1988. It defines a set of framing standards that dictate how bytes are transmitted across links together with ways of multiplexing existing transmission line frames (e.g., DS-1, DS-3) into SONET.

The SONET format is developed to define a synchronous optical hierarchy that is flexible enough to carry many different types of payload. A byte-interleaved multiplexing scheme is adopted in SONET with a basic rate of 51.84 Mbps. A family of standard rates and formats based on multiples of the 51.84 Mbps STS-1 (Synchronous Transport Signal-level 1) signal are defined at a rate of *n* times 51.84 Mbps, where *n* is an integer. See Table 5.1 for the SONET/SDH hierarchy. The basic signal at 51.84 Mbps is called STS-1. The direct optical conversion of the STS-1 signal is called OC-1 (optical carrier –level 1). The higher level signals are denoted by STS-N and OC-N, where N is an integer. Refer to Bellcore GR-253-CORE for more information.

The term line, section, and path are used to delineate various paths of the transmission network that interconnect the SONET elements. Figure 5.1 illustrates the SONET line, section, and path.

SONET Frame Structure

The SONET frame consists of the following two components:

1. Synchronous payload envelope;
2. Transport overhead.

TABLE 5.1. SONET/SDH Hierarchy

Optical carrier level	SDH level ITU	SONET level ANSI (electrical)	Data rate
OC-1		STS-1	51.84 Mbps
OC-3	STM-1	STS-3	155.52 Mbps
OC-12	STM-4	STS-12	622.08 Mbps
OC-24	STM-8	STS-24	1.244 Gbps
OC-48	STM-16	STS-48	2.488 Gbps
OC-N	STM-N/3	STS-N	N x 51.84 Mbps

OC: optical carrier
STS: synchronous transport signal
STM: synchronous transport module

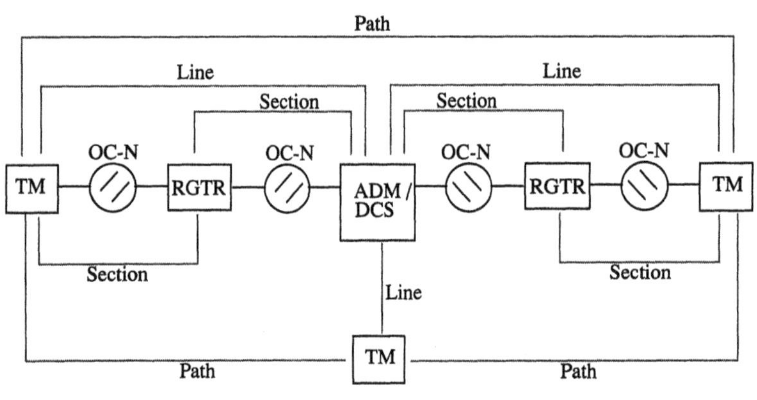

DCS: digital cross-connect system
ADM: add-drop multiplex
RGTR: regenerator
TM: terminal multiplex

FIGURE 5.1. SONET line, section, and path. (*After:* GR-253-CORE, 1995.)

Synchronous Payload Envelope

The payload information is carried over the synchronous payload envelope (SPE) data structure. The SPE contains the following two items:

1. Path overhead;
2. Payload.

The path overhead (POH) contains reassembly-specific information of the original data and the payload user data that are being transmitted. The user data may be ATM cells, VT-mapped DS-1s, or any other form of information properly mapped into SONET.

Transport Overhead

The transport overhead (TOH) ensures reliable communication among network equipment for certain functions. Some of these are error monitoring, network operations, and administration. This overhead is subdivided into two parts:

1. Section overhead;
2. Line overhead.

The section overhead (SOH) deals with the section terminating equipment, and the line overhead (LOH) deals with the line terminating equipment.

The SONET hierarchy is built on four basic formats:

1. Frame structure of the STS-1;
2. Frame structure of the STS-N;
3. Frame structure of the STS-Nc;
4. Virtual tributary structures.

Frame Structure of the STS-1

The basic SONET frame structure is an STS-1 format. The data rate for STS-1 is 51.840 Mbps; that is, 810 octets transmitted every 125 μs. In the digital voice network, the minimal sampling frequency for voice is 8,000 Hz (one sample every 125 μs) because of the Nyquist criteria, which is that the minimal sampling frequency must be equal to or greater than twice the bandwidth. The bandwidth of the human voice is between 50 to 4,000 Hz. See Figure 5.2 for the SONET frame structure (STS-1).

See Figure 5.3 for the SONET STS-1 frame structure showing overhead locations. The SONET SPE (POH and payload) floats within the SONET frame.

The detailed description of the overhead octets can be found in Bellcore document GR-253-CORE.

See Table 5.2 for the brief descriptions of the SONET STS-1 frame SOH octets.

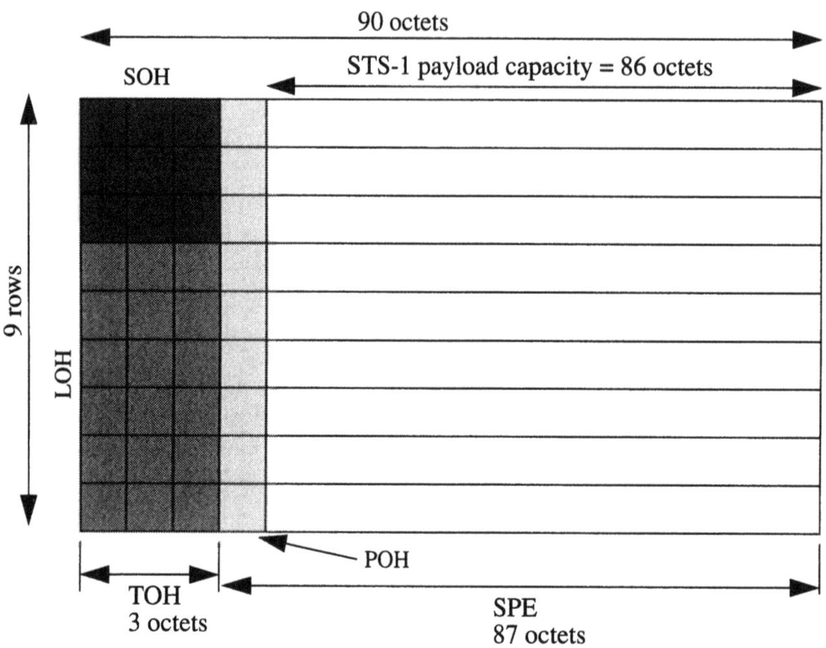

FIGURE 5.2. SONET frame structure (STS-1).

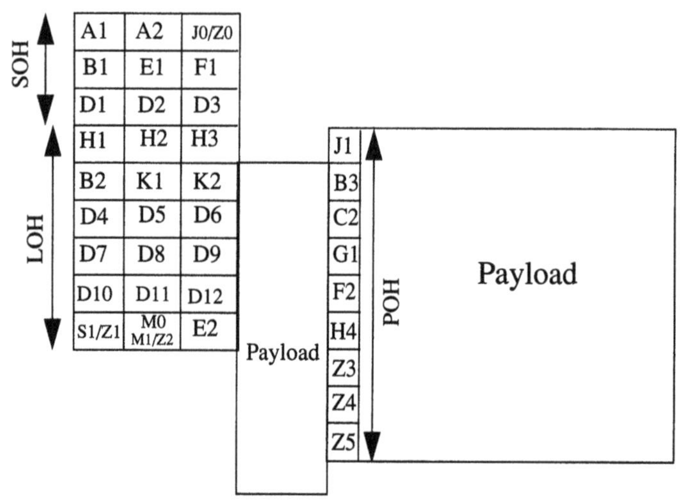

FIGURE 5.3. SONET STS-1 frame structure showing overhead locations.

TABLE 5.2. SONET STS-1 Frame SOH Octets

Overhead	Coding	Description
A1, A2	A1: 11110110, A2: 00101000	Framing octets
J0/Z0	J0: 00000001- 1st Z0: 00000010 2nd Z0:00000011	Section trace (J0)/section growth (Z0)
B1	Section BIP-8	Section error monitoring (previous STS-1)
E1	Orderwire	Section orderwire for communication between regenerators, hubs, remote terminal locations
F1	Section user channel	Section user channel, which is available for use by the network provider
D1, D2, D3	Section data communication channel	Three octets operate as one 192-kbps message-based ch for alarms, maintenance, control, monitoring, administering

Example of BIP-8

The bit interleaved parity-8 (BIP-8) code is calculated over entire packet using even or odd parity in the following manner:

- BIP-8 (bit 8): parity calculated over bit 8 of every octet of the entire packet;
- BIP-8 (bit 7): parity calculated over bit 7 of every octet of the entire packet;
- BIP-8 (bit 6): parity calculated over bit 6 of every octet of the entire packet;
- BIP-8 (bit 5): parity calculated over bit 5 of every octet of the entire packet;
- BIP-8 (bit 4): parity calculated over bit 4 of every octet of the entire packet;
- BIP-8 (bit 3): parity calculated over bit 3 of every octet of the entire packet;
- BIP-8 (bit 2): parity calculated over bit 2 of every octet of the entire packet;
- BIP-8 (bit 1): parity calculated over bit 1 of every octet of the entire packet.

See Table 5.3 for the brief descriptions of the SONET STS-1 frame LOH octets

See Table 5.4 for the brief descriptions of the SONET STS-1 frame POH octets.

Table 5.5 shows the STS path signal label assignments.

Table 5.6 shows the STS path status octet (G1).

Frame Structure of the STS-N

STS-N formats allow the transportation of *n* number of STS-1 payloads. The STS-N is formed by byte-interleaving STS-1 and STS-M (M<N) modules. The TOH of the individual STS-1 and STS-M modules are frame aligned before interleaving,

TABLE 5.3. SONET STS-1 Frame LOH Octets

Overhead	Coding	Description
H1, H2	STS payload pointer	Two octets are used to indicate the 1st octet of the STS SPE.
H3	Pointer action	Pointer action octet is allocated for SPE frequency justification purposes.
B2	Line BIP-8	Line error monitoring (previous STS-1).
K1, K2	APS channel	Used for automatic protection switching (APS) signaling between LTE; K2 also used to detect AIS-L and RDI-L signals.
D1 through D12	Line data communication channel	Nine octets operate as one 576-kbps message-based ch for alarms, maintenance, control, monitoring, administering.
S1	Synchronization status 1st STS-1 of an STS-N	Bits 5 through 8 of S1 are allocated to convey the synchronization status of the NE. Bits 1 through 4 of S1 are not defined.
Z1	Growth 2nd through Nth of an STS-N	Future use.
M0	STS-1 REI-L Only for the STS-1 in an OC-1	Bits 5 through 8 of the M0 are allocated for a line remote error indication (REI-L) function. Bits 1 through 4 of the M0 are not defined.
M1	STS-N REI-L 3rd STS-1 of an STS-N	M1 octet is set to indicate (to the upstream LTE) the count of the interleaved-bit block errors that it has detected using the line BIP-8 (B2) octets.
Z2	Growth	Future use.
E2	Orderwire	Line orderwire for communication between line entities.

LTE: line terminating equipment
AIS-L: alarm indication signal-line
RDI-L: remote defect indication-line
REI-L: remote error indication-line
BIP-8: bit interleaved parity-8

but the associated STS SPEs are not required to be aligned because each STS-1 uses its own payload pointer to indicate the location of the SPE.

An STS-3 SONET frame structure is shown in Figure 5.4. The STS-3 frame is very similar to the STS-1 frame. The STS-3 is formed by byte-interleaving STS-1 modules. The data rate is 155.52 Mbps; that is, 2,430 octets transmitted every 125 μs.

Frame Structure of the STS-Nc

STS-Nc formats allow the transportation of larger than N number of STS-1 payloads by concatenation. To accommodate such payloads, an STS-Nc module is

TABLE 5.4. SONET STS-1 Frame POH Octets

Overhead	Coding	Description
J1	STS path trace	Repetitive 64-octet message to verify its continued connection.
B3	STS path BIP-8	Path error monitoring (previous STS-1).
C2	STS path signal label	Indicates the content of the STS SPE, including the status of the mapped payloads.
G1	Path status	Conveys the path terminating status and performance back to the originating STS PTE.
H4	Indicator VT-structured STS-1 SPEs, ATM, and DQDB mapping	Multiframe indicator for VT-structured STS-1 SPEs and carry DQDB link status signal.
F2	Path user channel	Path user channel, which is available for use by the STS path terminating NEs.
Z3, Z4	STS path growth	Undefined.
Z5	Tandem connection	Used for tandem connection maintenance and the path data channel.

PTE: path terminating equipment
VT: virtual tributaries
DQDB: distributed queue dual bus
BIP-8: bit interleaved parity-8
NE: network element

TABLE 5.5. STS Path Signal Label (C2) Assignments

Code (Hex)	Content of STS SPE
00	Unequipped
01	Equipped—nonspecific payload
02	VT-structured STS-1 SPE
03	Locked VT mode
04	Asynchronous mapping for DS-3
12	Asynchronous mapping for DS4NA
13	Mapping for ATM
14	Mapping for DQDB
15	Asynchronous mapping for FDDI

Note: STS path signal label assignments for signals with payload defects are not shown here.

PTE: path-terminating equipment
VT: virtual tributaries
DQDB: distributed queue dual bus
FDDI: fiber distributed data interface

TABLE 5.6. STS Path Status Octet (G1)

REI-P (4 bits)	RDI-P (3 bits)	Undefined

REI-P coding:

0 through 8: error counts from 0 to 8
9 through 15: error counts = 0

RDI-P: remote defect indication-path
REI-P: remote error indication-path

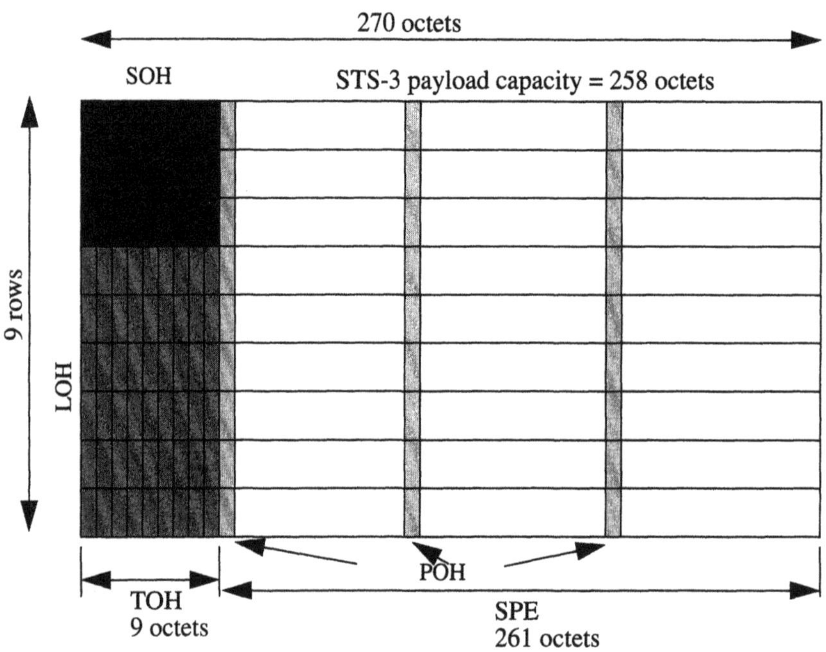

FIGURE 5.4. SONET frame structure (STS-3).

formed by linking N constituent STS-1s together in fixed-phase alignment. This payload is then mapped into the resulting STS-Nc SPE for transport.

An STS-Nc SPE consists of N x 783 octets (N x 87 columns by 9 rows). Only one set of POH is required in the STS-Nc SPE.

The STS-3c is similar to the STS-3, but the concatenated part is obtained by removing the POH for two of the frames. Three STS-1 SPEs are concatenated to form

STS-3c SPE. Therefore, the payload capacity and TOH are three times larger than an STS-1 frame, while only one POH exists. See Figure 5.5 for the SONET frame structure (STS-3c).

The STS-3c is the choice for ATM today. The data rate for STS-3c is 155.52 Mbps. ninety octets out of 2,430 octets are used by the TOH and POH. The remaining 2340 octets (96.3% of the SONET frame) are used for ATM payload. Therefore, the amount of SONET payload available in an STS-3c frame is 149.76 Mbps.

$$\frac{(2{,}430 \text{ octets} - 90 \text{ octets})155.52 \text{ Mbps}}{2{,}430 \text{ octets}} \approx 149.76 \text{ Mbps}$$

$$\frac{2{,}430 \text{ octets} - 90 \text{ octets}}{53 \text{ octets}} \approx 44.15 \text{ ATM cells per frame}$$

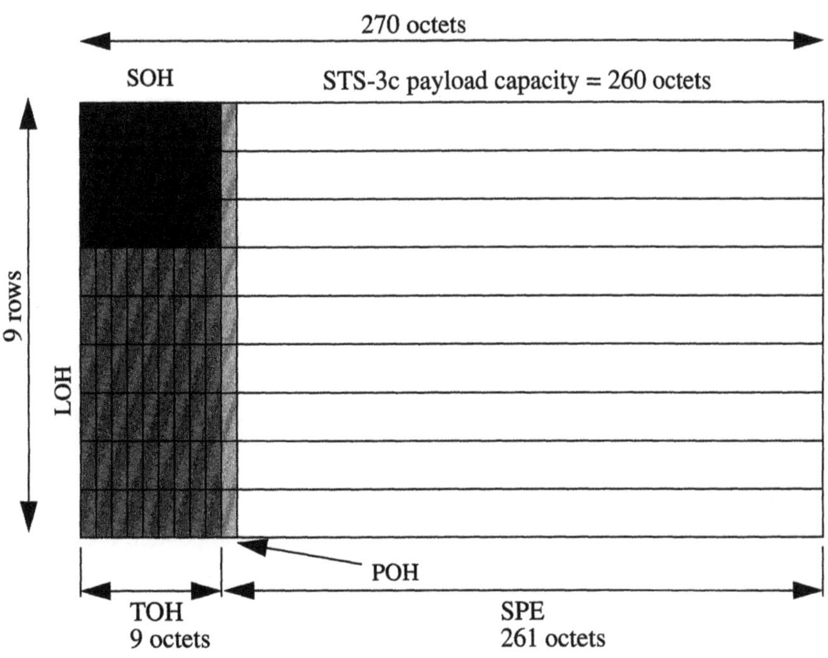

Payload rate: (9 rows x 260 columns x 8 bits) / 125 μs = 149.76 Mbps

FIGURE 5.5. SONET frame structure (STS-3c).

Approximately 44 ATM cells fit into the STS-3c frame because ATM cells are 53 octets long. ATM cells can cross SONET frame boundaries.

The STS-3c physical layer interface, ATM cell mapping, frame structure, and a design example are described later in this chapter.

Virtual Tributary Structures

The virtual tributary (VT) structure is designed to transport and switch sub–STS-1 rate payloads. The VT structure consists of a VT pointer to locate the start of VT SPE. The VT SPE is composed of VT POH and payload. There are four sizes of VTs:

1. VT1.5 (1.728 Mbps);
2. VT2 (2.304 Mbps);
3. VT3 (3.456 Mbps);
4. VT6 (6.912 Mbps).

The VT-structured STS-1 SPE is divided into seven VT groups. Each VT group occupies 12 columns of the 87-column STS-1 SPE and may contain four VT1.5s, three VT2s, two VT3s, or one VT6. A VT group contains only one size of VTs; however, a different VT size is allowed for each VT group in an STS-1 SPE.

Figure 5.6 shows VT sizes for VT1.5 and VT2, and Figure 5.7 shows VT sizes for VT3 and VT6.

An example of mapping VT1.5 into STS-1 SPE is shown in Figure 5.8 where all VT groups contain VT1.5s.

In addition to the division of VTs into VT groups, four consecutive 125 μs frames of the VT-structured STS-1 SPE are organized into a 500 μs VT superframe. The phase of the VT superframe is indicated by the H4 octet in the STS-1 POH. The VT superframe contains V1, V2, V3, and V4 octets, which are described below.

V1 and V2: VT Payload Pointer

These octets of the VT superframe are used as the VT payload pointer. It provides flexible and dynamic alignment of the VT SPE within the VT envelope capacity.

V3: VT Pointer Action Octet

The VT pointer action octet is used for positive or negative movement of the payload. If an increment is detected in the payload pointer, then the octet following V3 is considered a positive stuff octet, and the current pointer value is incremented by one. If a decrement is detected in the payload pointer, then the V3 octet is considered a negative stuff octet, and the current pointer value is decremented by one.

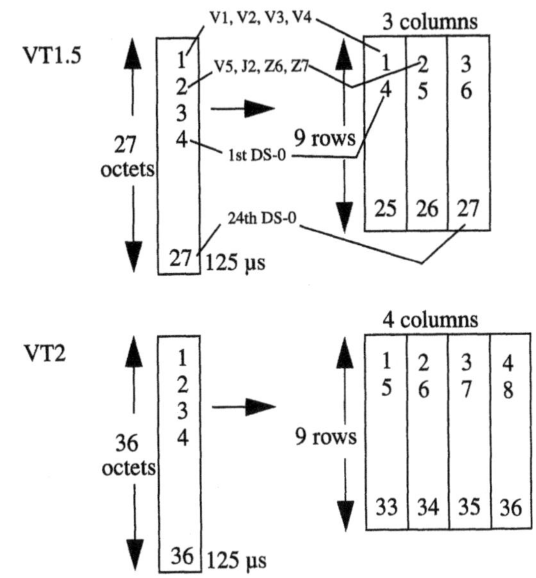

FIGURE 5.6. VT sizes (VT1.5 and VT2).

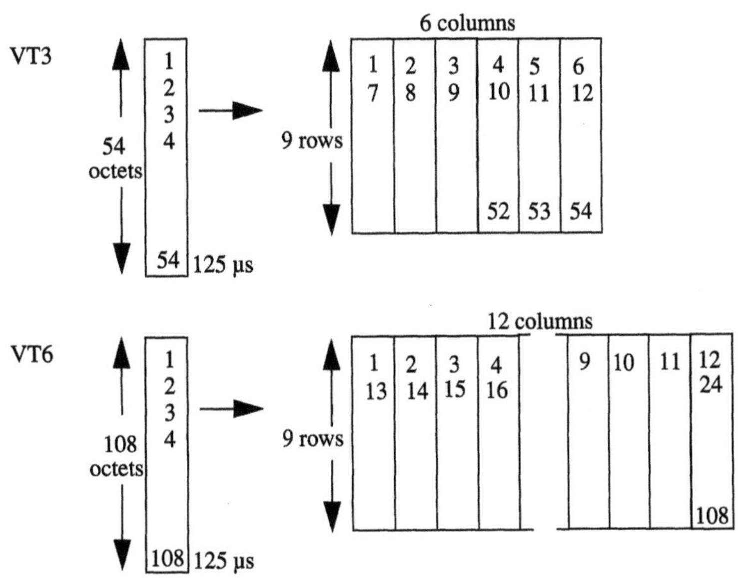

FIGURE 5.7. VT sizes (VT3 and VT6).

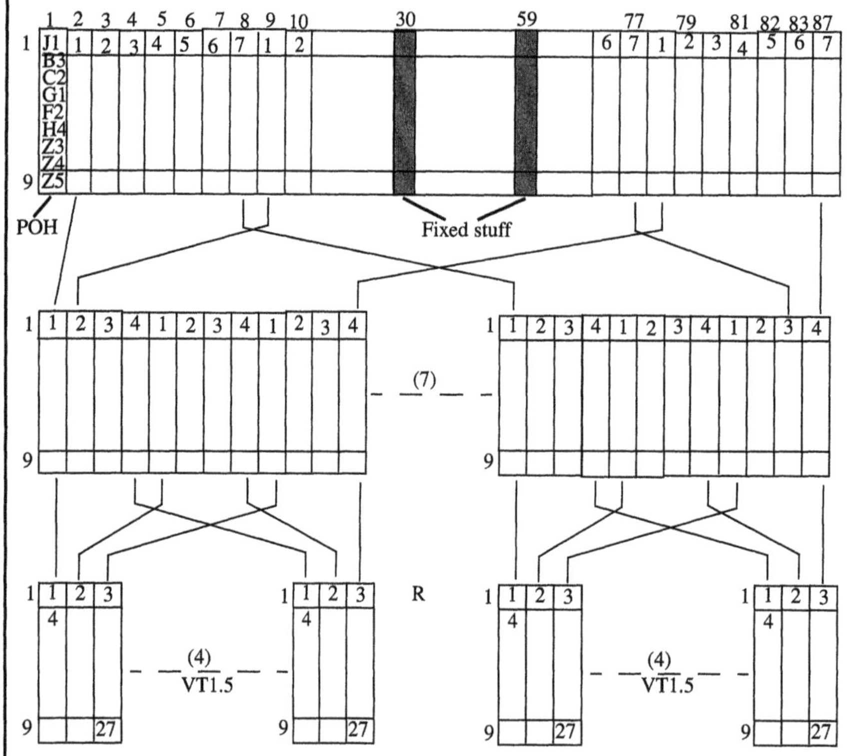

Note: Each block represents an octet.

FIGURE 5.8. Example of VT1.5-structured STS-1 SPE.

V4: Unused

Not defined.

VT SPE

Each VT SPE contains four octets of VT POH (V5, J2, Z6, and Z7), and the VT payload capacity, which is different for each VT size. Figure 5.9 shows the VT superframe and envelope capacity, and Figure 5.10 shows the VT SPE and payload capacity.

VT Path Overhead

Four octets (V5, J2, Z6, and Z7) are defined as VT POH. The first octet location pointed by the VT payload pointer is the V5 octet, while the J2, Z6, and Z7 octets are located at the corresponding locations in the subsequent 125 μs frames of the VT superframe.

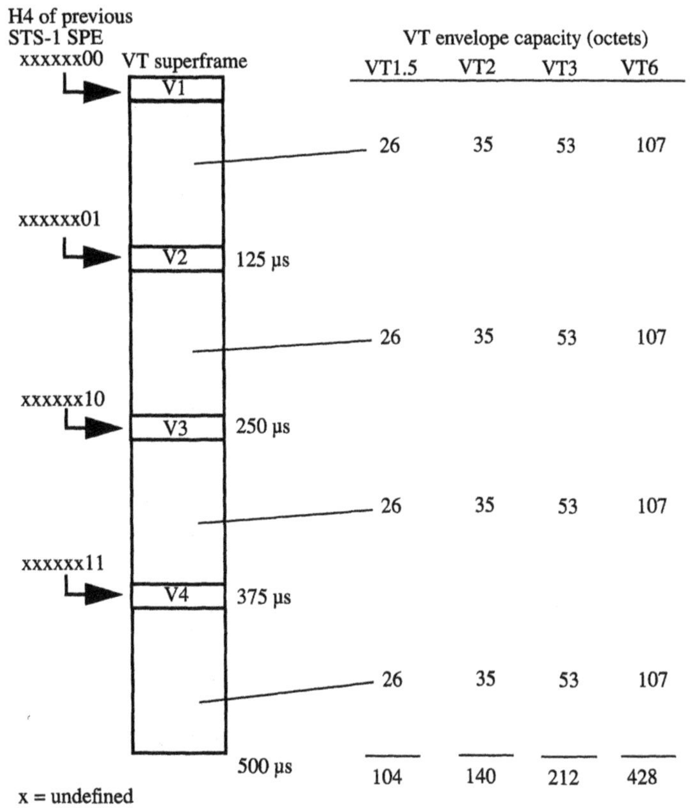

FIGURE 5.9. VT superframe and envelope capacity.

Table 5.7 shows the VT path overhead octet (V5). The J2 (VT path trace) octet is reserved for VT path trace function. The Z6 (VT path growth) octet is allocated for future growth. Bits 5 through 7 of the Z7 (VT path growth) octet (along with the bit 8 of the V5 octet) are allocated for an RDI-V signal.

Sub-STS-1 Mapping

Payloads below the DS-3 rate are transported in a VT structure.

Byte-Synchronous Mapping for DS-1

A byte-synchronous mapping of a DS-1 into the payload of a VT1.5 SPE allows downstream SONET NEs to locate and access twenty-four DS-0 channels directly. This allows twenty-four DS-0 channels to be transported without a DS-1

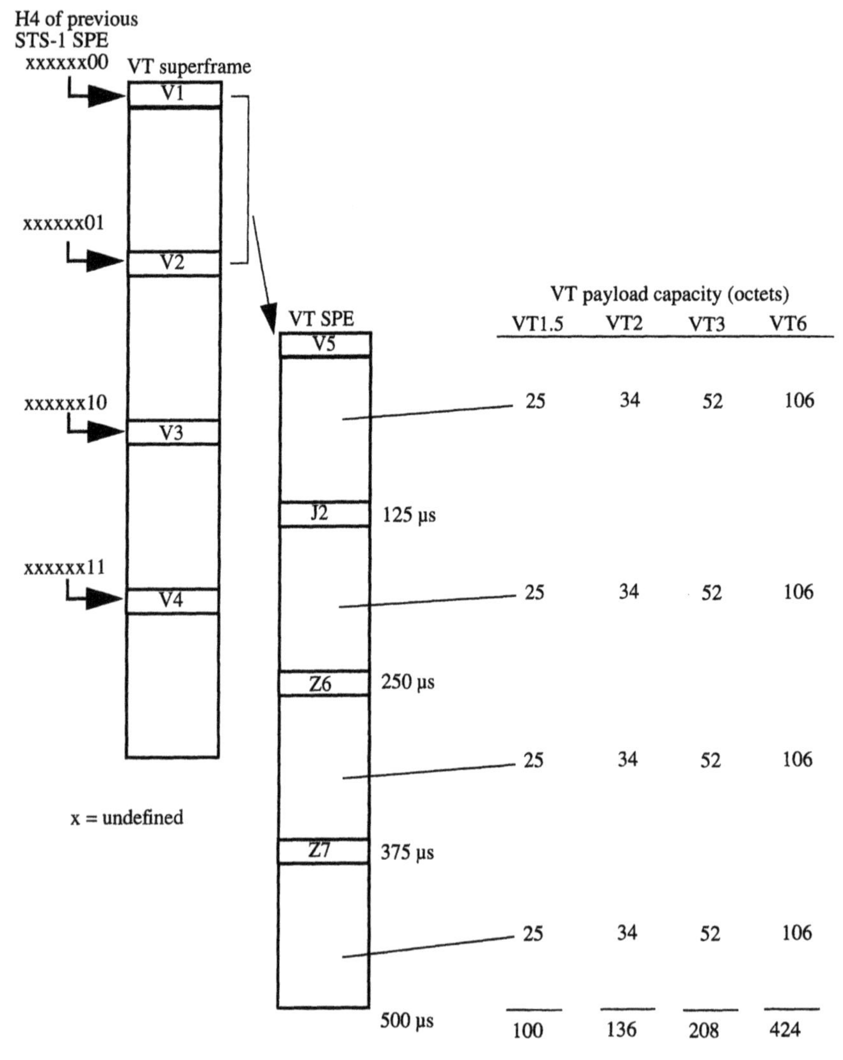

FIGURE 5.10. VT SPE and payload capacity.

interface. Figure 5.11 shows the byte-synchronous mapping for a DS-1 (or twenty-four DS-0s) into a VT1.5 SPE.

Asynchronous Mapping for DS-1

An asynchronous mapping of a DS-1 into the payload of a VT1.5 SPE allows clear-channel transportation of the DS-1 signal. Figure 5.12 shows the asynchronous mapping for a DS-1 into a VT1.5 SPE.

TABLE 5.7. VT POH Octet (V5)

BIP-2 2 bits	REI-V 1 bit	RFI-V 1 bit	Signal Label 3 bits	RDI-V 1 bit

REI-V coding
0: 0 error
1: 1 or 2 errors

Signal label coding Note: VT1.5 signal label assignments only

000: unequipped
001: equipped-nonspecific payload
010: asynchronous mapping for DS-1
011: bit-synchronous mapping for DS-1
100: byte-synchronous mapping for DS-1

REI-V: remote error indication (VT path)
RFI-V: remote failure indication (VT path)
RDI-V: remote defect indication (VT path)
BIP-2: bit-interleaved parity-2

The asynchronous DS-1 mapping contains 771 = [(24 x 4 x 8) + 3i] information (I) bits, 6 stuff control (C) bits, 2 stuff opportunity (S) bits, and 8 overhead communication channel (O) bits in each VT1.5 SPE.

C1C1C1 = 0 0 0 indicates that S1 is an information bit, while C1C1C1 = 1 1 1 indicates that S1 is a stuff bit (used as filler). C2C2C2 bits control the S2 bit in the same way.

STS-1 Mapping

Asynchronous Mapping for DS-3

An asynchronous mapping for a DS-3 into the payload of an STS-1 SPE allows clear channel transportation of the DS-3 signal. Figure 5.13 shows the asynchronous mapping for a DS-3 into an STS-1 SPE.

The asynchronous DS-3 mapping contains nine subframes every 125 μs. Each subframe contains 621 = [(25 x 3 x 8) + (2 x I) +5] information (i) bits, five stuff control (c) bits, one stuff opportunity (s) bit, and two overhead communication channel (o) bits.

c c c c = 0 0 0 0 indicates that s is an information bit, while c c c c = 1 1 1 1 indicates that s is a stuff bit.

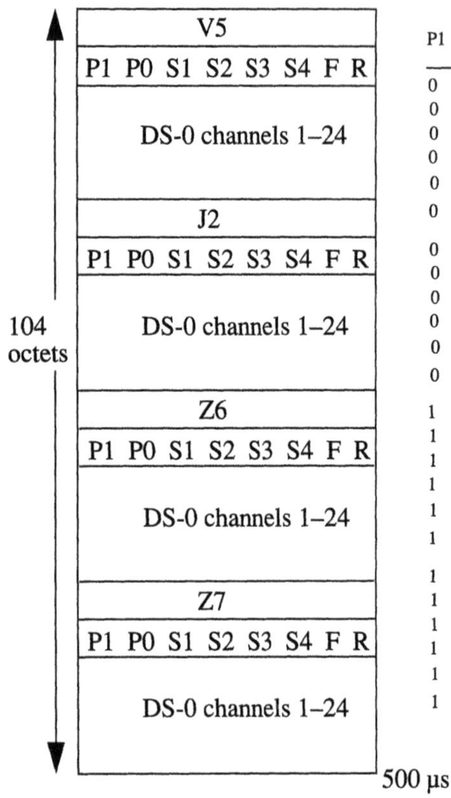

P1	P0	S1	S2	S3	S4	SF F	ESF F
0	0	A1	A2	A3	A4	F1	M1
0	0	A5	A6	A7	A8	S1	C1
0	0	A9	A10	A11	A12	F2	M2
0	0	A13	A14	A15	A16	S2	F1
0	0	A17	A18	A19	A20	F3	M3
0	0	A21	A22	A23	A24	S3	C2
0	1	B1	B2	B3	B4	F4	M4
0	1	B5	B6	B7	B8	S4	F2
0	1	B9	B10	B11	B12	F5	M5
0	1	B13	B14	B15	B16	S5	C3
0	1	B17	B18	B19	B20	F6	M6
0	1	B21	B22	B23	B24	S6	F3
1	0	C1	C2	C3	C4	F1	M7
1	0	C5	C6	C7	CA8	S1	C4
1	0	C9	C10	C11	C12	F2	M8
1	0	C13	C14	C15	C16	S2	F4
1	0	C17	C18	C19	C20	F3	M9
1	0	C21	C22	C23	C24	S3	C5
1	1	D1	D2	D3	D4	F4	M10
1	1	D5	D6	D7	D8	S4	F5
1	1	D9	D10	D11	D12	F5	M11
1	1	D13	D14	D15	D16	S5	C6
1	1	D17	D18	D19	D20	F6	M12
1	1	D21	D22	D23	D24	S6	F6

PO P1 : phase indicator
S1, S2, S3, S4: signaling bits
F: DS-1 framing bit
R: fixed stuff (used as filler)

FIGURE 5.11. Byte-synchronous mapping for DS-1.

ATM Mapping for B-ISDN Application

ATM cells consist of a 5-octet cell header and a 48-octet payload. ATM cells are mapped into the STS-1 payload by aligning the octet structure of every cell with the octet structure of the STS-1 SPE. The entire STS-1 payload is filled with ATM cells, yielding a transfer capacity for ATM cells 49.54 Mbps.

Because the STS-1 payload is not an integer multiple of the 53-octet ATM cell length, some cells cross the STS-1 SPE boundary. Figure 5.14 shows the ATM cells are mapped into the payload of the STS-1 SPE. The first ATM cell in the STS-1 SPE is pointed to by the H4 octet, which is the number of octets from the J1 octet.

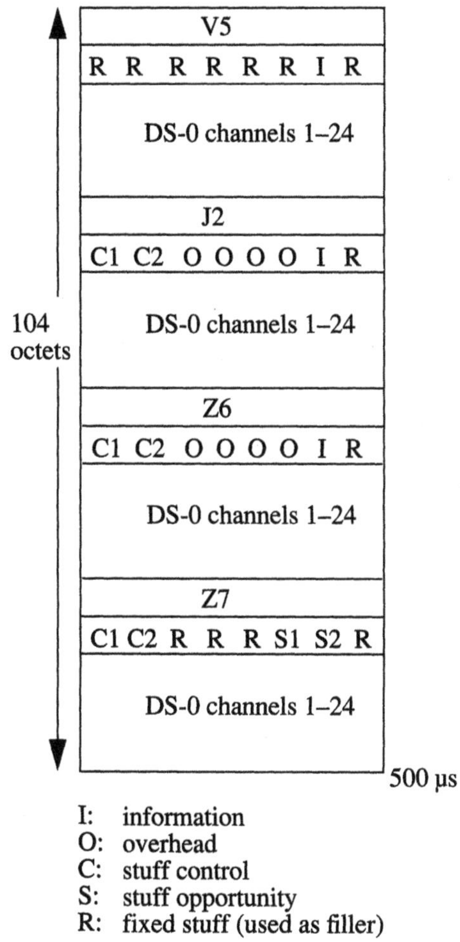

FIGURE 5.12. Asynchronous mapping for DS1.

Frame Structure of the STS-3c

The SONET STS-3c frame format is used at the 155.52-Mbps B-ISDN user-network interface. The STS-3c formats allow the transportation of larger than three STS-1 payloads by concatenation. To accommodate such payloads an STS-3c module is formed by linking three constituent STS-1s together in fixed-phase alignment. This payload is then mapped into the resulting STS-3c SPE for transport.

An STS-3c SPE consists of 3 x 783 octets (3 x 87 columns by 9 rows). Three STS-1 SPEs are concatenated to form STS-3c SPE. Therefore, the payload capacity and

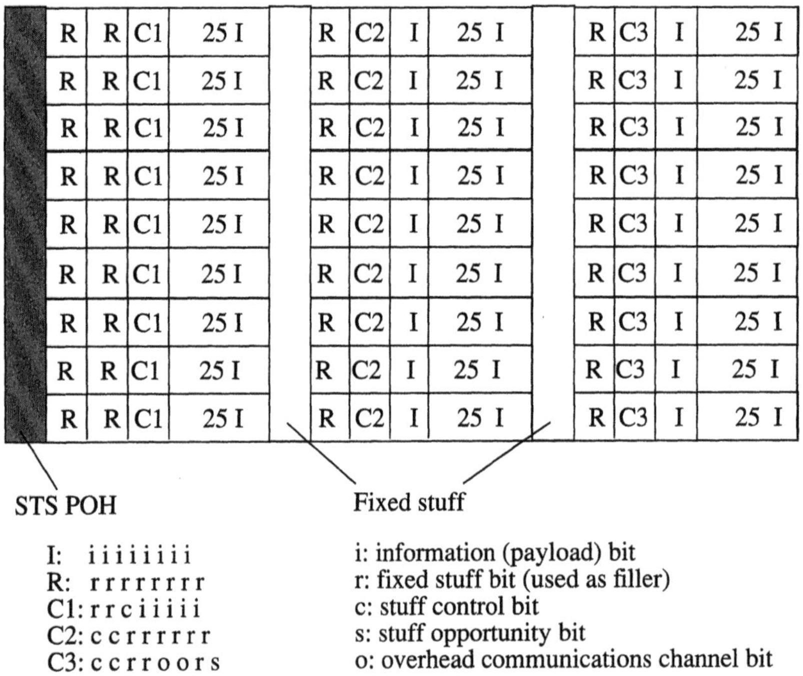

FIGURE 5.13. Asynchronous mapping for DS-3 payload.

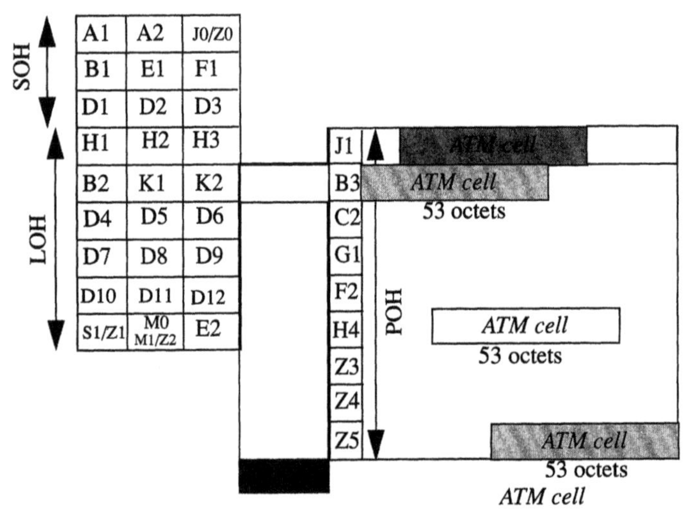

FIGURE 5.14. Mapping for ATM payload.

TOH are three times larger than an STS-1 frame, while only one POH exists. See Figure 5.15 for the SONET STS-3c frame structure at the UNI.

A detailed description of the overhead octets can be found in GR-253-CORE.

Table 5.8 shows the STS-3c frame section overhead octets.

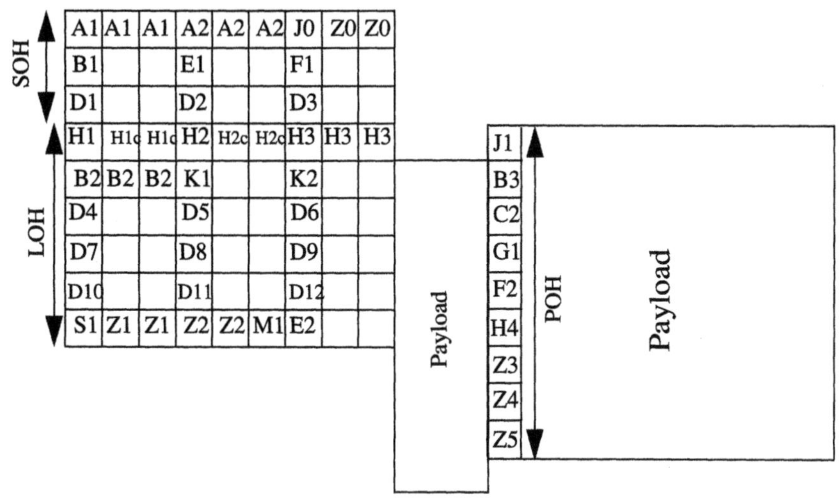

FIGURE 5.15. SONET STS-3c frame structure at the UNI.

TABLE 5.8. SONET STS-3c Frame SOH Octets

Overhead	Coding	Description
A1, A2	A1: 11110110, A2: 00101000	Framing octets
J0, Z0, Z0	Jo: 00000001- 1st Z0: 00000010 2nd Z0:00000011	Section trace (J0)/section growth (Z0)
B1	Section BIP-8 (bit interleaved parity-8)	Section error monitoring (previous STS-1)
E1	Orderwire	Section orderwire for communication between regenerators, hubs, remote terminal locations
F1	Section user channel	Section user ch, which is available for use by the network provider
D1, D2, D3	Section data communication channel	Three octets operate as one 192-kbps message-based ch for alarms, maintenance, control, monitoring, administering

STS-3c Overhead Octet Description

A1, A2: Frame Alignment Pattern

The A1, A2 octets are coded as F6F6F6282828H on the transmit side. On the receive side, the UNI searches the SONET frame alignment pattern. When the pattern has been detected for two consecutive frames, the UNI declares in-frame.

When errors are detected in the pattern for four consecutive frames, the UNI declares out-of-frame.

J0, Z0, Z0: Section Trace (J0)/ Section Growth (Z0)

The J0, Z0, Z0 octets are coded as 010203H. The UNI may use these octets as an additional frame alignment pattern (in addition to Al and A2). Normally, these bytes are ignored by the receiver.

B1: Section-Bit Interleaved Parity

This octet contains a bit-interleaved parity-8 (BIP-8) calculated across the entire SONET frame (810 x 3 = 2,430 octets). The current frame contains the BIP result calculated for the previous frame. The B1 value is calculated based on even parity. On the receiving side, the BIP-8 is calculated for the current frame and stored. The B1 octet contained in the subsequent frame is extracted and compared with the calculated value. Differences between the two values provide an indication of the bit-error rate across a SONET section.

E1: Section Orderwire

The E1 octet is coded with all zeros for the UNI. This is the section orderwire, and it contains a 64-Kbps PCM voice channel for communication across a SONET section.

F1: Section User Channel

The F1 octet is coded with all zeros for the UNI. This is the section user channel and it contains a 64-Kbps data link channel for proprietary use by the network provider for the NNI.

D1–D3: Section Data Communications Channel

These octets are coded with all ones for the UNI. These are the section data communications channel (DCC) and contain a 192-Kbps data link channel for network operation, administration, maintenance, and provisioning for the NNI.

Table 5.9 shows the STS-3c frame LOH octets.

TABLE 5.9. SONET STS-3c Frame LOH Octets

Overhead	Coding	Description
H1, H2	STS payload pointer	Two octets are used to indicate the first octet of the STS SPE.
H1c, H2c	Concatenation indication Specifies only one POH in STS-Nc SPE	H1c = 1001xx11 H2c = 11111111
H3	Pointer action	Pointer action octet is allocated for SPE frequency justification purposes.
B2	Line BIP-8	Line error monitoring (previous STS-1).
K1, K2	APS channel	Used for automatic protection switching (APS) signaling between LTE; K2 also used to detect AIS-L and RDI-L signals.
D1 through D12	Line data communication channel	Nine octets operate as one 576-kbps message-based ch for alarms, maintenance, control, monitoring, administering.
S1	Synchronization status first STS-1 of an STS-N	Bits 5 through 8 of S1 are allocated to convey the synchronization status of the NE. Bits 1 through 4 of S1 are not defined.
Z1	Growth 2nd through Nth of an STS-N	Future use.
M1	STS-N REI-L 3rd STS-1 of an STS-N	M1 octet is set to indicate (to the upstream LTE) the count of the interleaved-bit block errors that it has detected using the line BIP-8 (B2) octets.
Z2	Growth	Future use.
E2	Orderwire	Line orderwire for communication between line entities. Future use.

LTE: line terminating equipment
AIS-L: alarm indication signal-line
RDI-L: remote defect indication-line
REI-L: remote error indication-line
BIP-8: bit interleaved parity-8

H1, H2, H3: Payload Pointer, Pointer Action

The payload pointer points to the first octet (J1) of the synchronous payload envelope (SPE). The payload pointer is used to accommodate the jitter that accumulates in the transmission system. The change in the pointer values causes the SPE to move (three octets at a time) within the SONET frame. The pointer action octets (H3) carry data when a negative pointer movement is inserted. The three SPE octets immediately following the H3 octets are stuffed when a positive pointer movement is inserted.

The fixed pointer value 600000H is coded at the UNI. With this value, the first byte of the synchronous payload envelope (J1) is located immediately following the H3 octets in the SONET frame.

In the receiving end, the payload pointer is interpreted to locate the J1 octet. Although the pointer is fixed when the ATM cells are originally inserted (the origin of the SONET path), the payload pointer changes as the path travels through a SONET network. The pointer interpretation rules are contained in GR-253-CORE.

H1c, H2c: Pointer Concatenation Indicators

These octets are coded with 9393FFFFH. These octets indicate that the SONET is structured as a single, contiguous payload envelope (STS-3c) instead of three payload envelopes, each with its own independent payload pointer (STS-3). These octets may be monitored by the receiver to distinguish between STS-3 and STS-3c SONET frames.

B2: Line Bit-Interleaved Parity

This octet contains a bit-interleaved parity-8 (BIP-8) calculated across the entire SONET frame (2,430 bytes) minus the 3 x 9 octets SOH portion of the TOH. The current frame contains the BIP result calculated for the previous frame. The B2 value is calculated based on even parity. On the receiving side, the BIP-8 is calculated for the current frame and stored. The B2 octet contained in the subsequent frame is extracted and compared with the calculated value. Differences between the two values provide an indication of the bit-error rate across a SONET line.

K1, K2: APS Channel

The APS channel implements a bit-oriented protocol used to initiate protection switching actions between two line-terminating network elements.

The K2 octet is also used to detect line alarm indication signal-line (AIS-L) and remote defect indication-line (RDI-L) signals.

D4–D12: Line Data Communications Channel

These octets are coded with all zeros for the UNI. The line DCC contains a 576-Kbps data link channel for network operation, administration, maintenance, and provisioning for NNI.

S1: Synchronization Status

Bits 5 through 8 of this octet are allocated to convey the synchronization status of the NE.

Z1: Line Growth

These octets are reserved for future definition.

Z2: Line Growth

These octets are reserved for future definition.

M1: STS-N REI-L

This octet is located in the third STS-1 of an STS-N and used for REI-L function. This octet is set to indicate (to the upstream line terminating equipment) the count of the interleaved-bit block errors that it has detected using the line BIP-8 (B2) octets.

E2: Line Orderwire

This octet is coded with all zeros for the UNI. The line orderwire contains a 64-Kbps PCM voice channel for crafts person communication across a SONET line for NNI.

Table 5.10 shows the STS-3c frame POH octets.

J1: Path Trace

This octet is unused in most private UNIs but may be required in a public UNI. When unused, the J1 octet is coded with all zeros and provides a 64-octet (ANSI) or a 16-octet (ITU-T) repeating pattern that allows the terminating path to verify its continued connection to the source for NNI.

B3: Path Bit-Interleaved Parity

This octet contains a bit-interleaved parity-8 (BIP-8) calculated across the entire SONET frame (2,430 bytes) minus the 9 x 9 octets of the TOH. The current frame

TABLE 5.10. SONET STS-3c Frame POH Octets

Overhead	Coding	Description
J1	STS path trace	Repetitive 64-octet message to verify its continued connection.
B3	STS path BIP-8	Path error monitoring (previous STS-1).
C2	STS path signal label	Indicates the content of the STS SPE, including the status of the mapped payloads.
G1	Path status	Conveys the path terminating status and performance back to the originating STS PTE.
H4	Indicator VT-structured STS-1 SPEs, ATM, and DQDB mapping	Multiframe indicator for VT-structured STS-1 SPEs and carry DQDB link status signal.
F2	Path user channel	Path user ch, which is available for use by the STS path terminating NEs.
Z3, Z4	STS path growth	Undefined.
Z5	Tandem connection	Used for tandem connection maintenance and the path data channel.

PTE: path terminating equipment
VT: virtual tributaries
DQDB: distributed queue dual bus
BIP-8: bit interleaved parity-8
NE: network element

contains the BIP result calculated for the previous frame. The B3 value is calculated based on even parity. On the receiving side, the BIP-8 is calculated for the current frame and stored. The B3 octet contained in the subsequent frame is extracted and compared with the calculated value. Differences between the two values provide an indication of the bit-error rate across a SONET path.

C2: Path Signal Label

The C2 octet indicates the content of the STS SPE, including the status of the mapped payloads. This octet is coded with 13 hexadecoimal (hex) to indicate the payload is filled with ATM cells.

G1: Path Status

The number of B3 errors detected at the near end is conveyed into the first four bit positions, which indicate the path REI-P. The REI-P field has nine legal values (0000–1000) indicating between zero and eight B3 errors. The code 1001 has been defined as a path RDI-P alarm. This code is sent when the near end is unable to delineate the receive cell stream. The fifth bit position is used to transmit a path yellow alarm. This bit is set when the receiver detects a SONET layer alarm (LOS, LOF, LOP, etc.). The receiver monitors the G1 octet for path status.

F2: Path User Channel

This octet is coded with all zeros for the UNI.

H4: Indicator

This octet is allocated for use as a mapping-specific indicator octet. This octet indicates the offset in octets to the next ATM cell boundary in the transmit stream. Cell-offset indicator ranges from 0 to 52. This pointer is generally redundant, as the header error check (HEC) mechanism for cell delineation provides the same function with better performance.

Z3, Z4, Z5: Path Growth

These octets are coded with all zeros. The Z5 octet is used to implement tandem path functionality for NNI.

SONET Protocol Stack

SONET has four layers of protocol stack. Each layer communicates to its peer equipment horizontally. The information is processed and vertically transferred to the other layer. The path layer accepts ATM payload and maps the services and

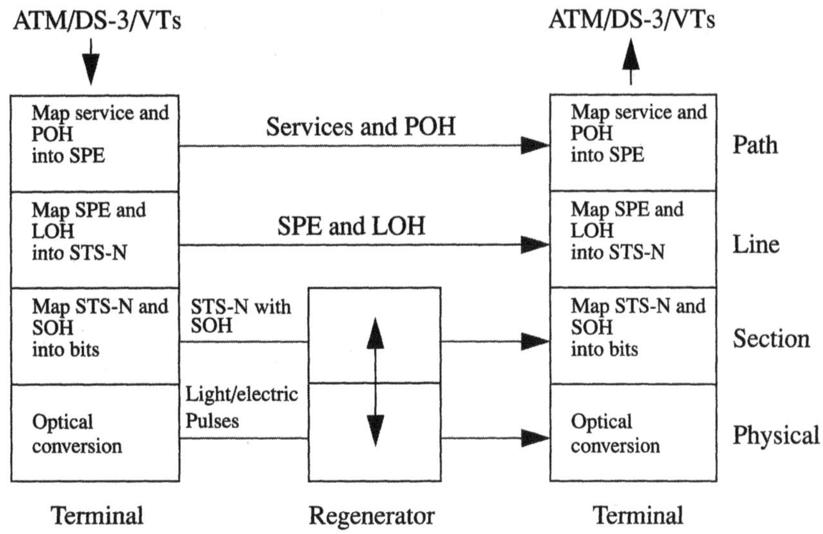

FIGURE 5.16. SONET layers.

POH into the SPE. It then passes the SPE to the line layer. The line layer maps the SPE and LOH into STS-N signals and hands it to the section layer. The section layer maps the STS-N and the SOH into pulses that are passed to the optical transceiver. See Figure 5.16 for the SONET layers.

SONET Fault Management

The fault management functions detect, isolate, and correct the failure conditions in the network. Fault management actions can be triggered by the following conditions:

- Incoming signal failures;
- Equipment failures;
- Detection or removal of AIS;
- Detection or removal of RDI signal.

The failures detected on the incoming signal can be listed as:

- Loss of Signal (LOS);
- Loss of Frame (LOF);

- Loss of Pointer (LOP);
- Signal label mismatch.

The flow diagram of the SONET maintenance interaction is shown in Figure 5.17.

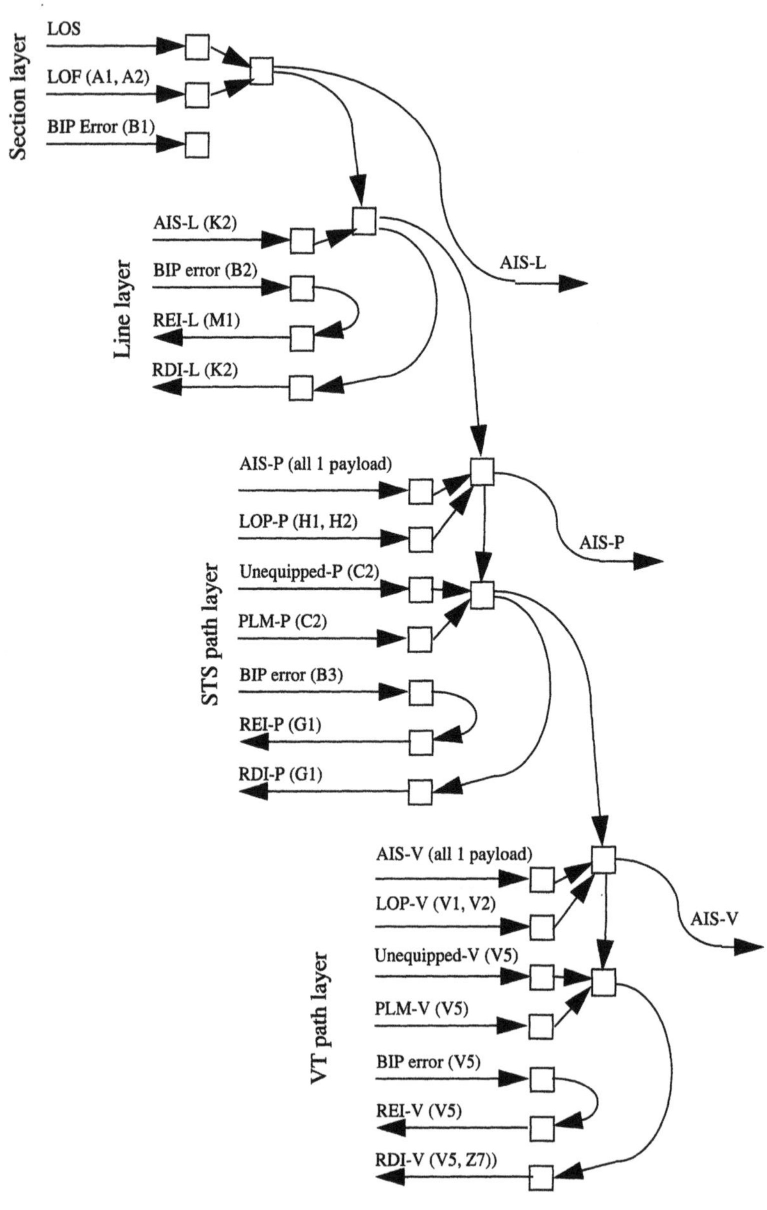

FIGURE 5.17. SONET maintenance interaction.

Example of STS-1 and OC-3 Signal Composition

An example of STS-1 and OC-3 signal composition is shown in Figure 5.18. This network element functions as STE, LTE, and PTE.

Path Level

STS-1 SPE is created by payload mapping (e.g., ATM cells into an STS-1 SPE) and adding the nine-octet POH. The coding of the following octets in the POH is listed below:

- BIP-8: Calculated using the previous STS-1 SPE;
- RDI-P: Set according to the SONET maintenance interaction;
- REI-P: Set upon received path BIP-8 (B3) error detection.

If the STS-1 SPE is unequipped, then it is indicated by inserting all zeros into the STS-1 SPE payload. Similarly, if the STS-1 SPE is in alarm, then it is indicated by inserting all ones into the STS-1 SPE payload to indicate AIS-P. The signal label (C2) is set to match the payload.

Line Level

An eighteen-octet LOH is added to the STS-1 SPE. The coding of the following octets in the LOH are listed below:

- BIP-8: Calculated using the previous STS-1 SPE plus the LOH;
- RDI-L: Set according to the SONET maintenance interaction;
- REI-L: Set upon received path BIP-8 (B2) error detection.

Pointer words are set to synchronize the STS-1 SPE. Three STS-1 SPEs are multiplexed to create STS-3 signal.

Section Level

A nine-octet SOH is added to the STS-3 SPE and the LOH. The coding of the following octets in the SOH are listed below:

- BIP-8: Calculated using the previous STS-1 SPE plus the TOH;
- REI-P: Set upon received path BIP-8 (B1) error detection.

Framing octets (A1, A2), section trace (J0), and section growth (Z0) are added after scrambling. The STS-3 signal is then converted into an optical signal.

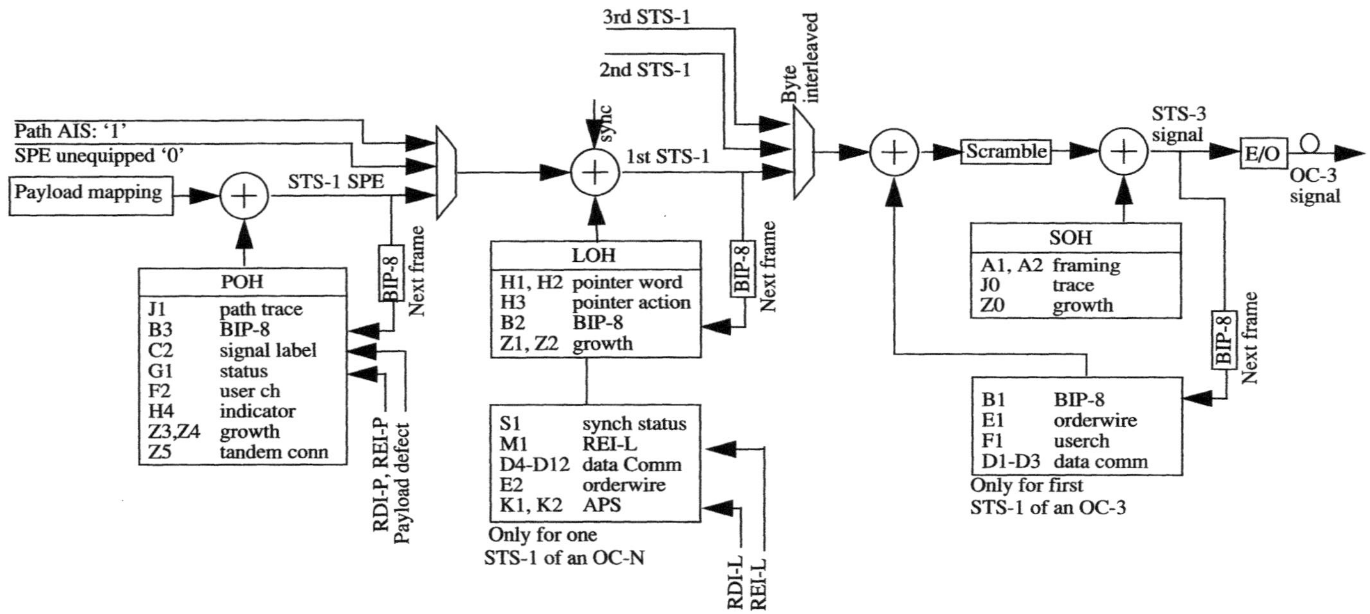

FIGURE 5.18. STS-1 and OC-3 signal composition. (*After:* GR-253-CORE, 1995.)

CHAPTER 6

SONET Pointer Processing

STS Payload Pointers

Pointers are defined in SONET at the STS-N and VT levels. The STS-N payload pointer provides a mechanism that allows the flexible and dynamic alignment of the STS-N SPE within the SONET envelope capacity. Dynamic alignment means that the STS-N SPE is allowed to float within the STS-N envelope capacity. Therefore, the pointer is able to accommodate phase differences of the STS-N SPE and the TOH. It also accommodates differences in the frame rates of the STS-N SPE and the TOH. Figure 6.1 shows that the STS-1 SPE will usually start in one frame and end in the following frame.

Refer to Bellcore GR-253-CORE for more information.

STS Pointer Value

The payload pointer octets (H1 and H2) of the LOH are viewed as one 16-bit word as shown in Figure 6.2. The four N-bits of the pointer word carry the new data flag, and the 10 I-bits and D-bits carry the pointer value. The two undefined bits are designated by dashes. The pointer value is a binary number with a range of 0 to 782 that indicates the offset between the pointer octets and the first octet of the SPE, which is the J1 octet. For example, a value of 0 indicates that the SPE starts in the octet position immediately following the H3 octet. The last column of this row is indicated by a pointer value of 86 (Figure 6.3). A pointer value of 87 denotes that the SPE starts at the octet position immediately following the K2 octet. A value of 782 would position the start of the SPE in the last column of the D3 octet (Figure 6.4).

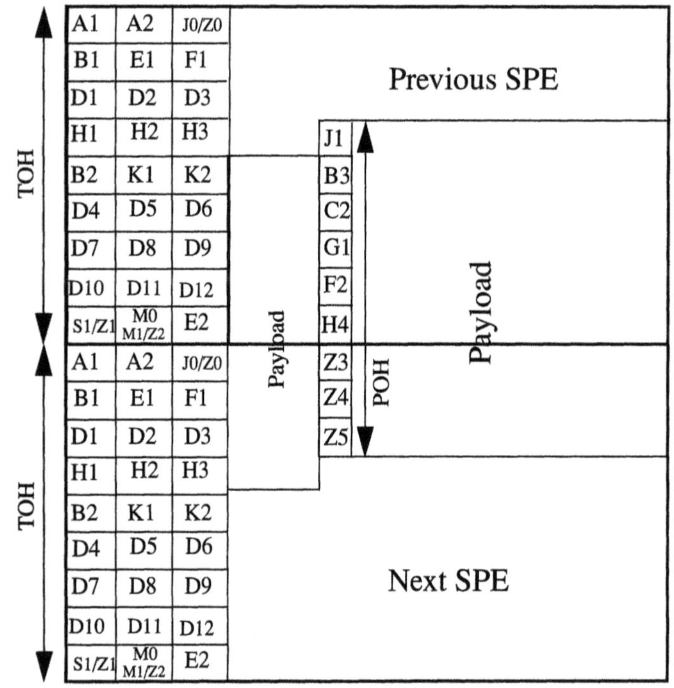

FIGURE 6.1. STS-1 frame.

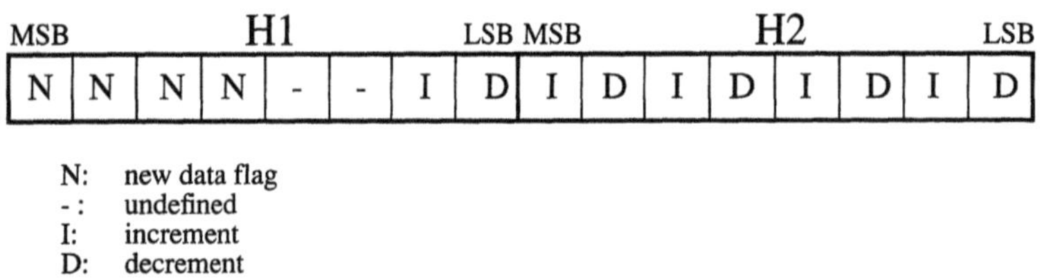

FIGURE 6.2. STS payload pointer (H1, H2) coding.

In the case of STS-Nc, it could be considered that there is a one-to-one correspondence, and that only the STS-Nc SPE octets that are associated with the first STS-1 of the STS-Nc are counted in determining the offset.

Alternatively, all of the octets in the STS-Nc SPE could be counted in determining the offset, and the NE could then transmit a pointer value equal to the offset divided by N. For example, a pointer value of 87 indicates that the SPE starts at the octet position immediately following the K2 octet, which is the same as the STS-1 SPE.

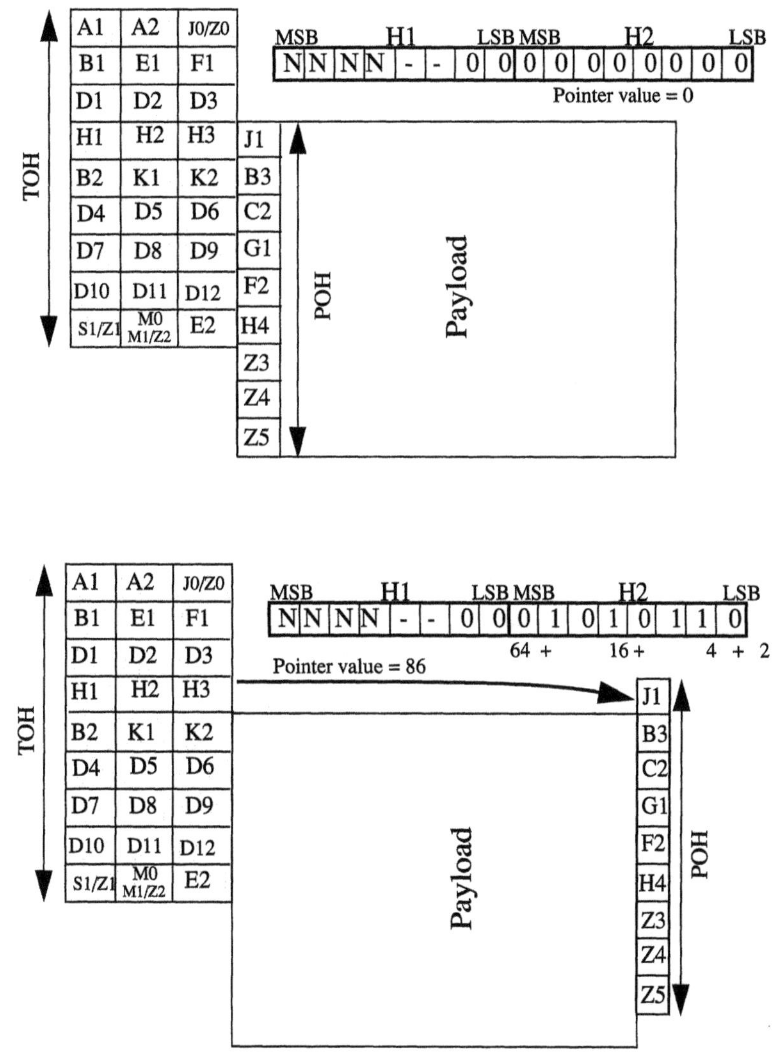

FIGURE 6.3. Example of the STS-1 pointer.

STS Frequency Justification

When there is a frequency offset between the frame rate of the TOH and that of the STS SPE, then the pointer value is incremented or decremented, as needed. This is accomplished by using a positive or a negative stuff octet.

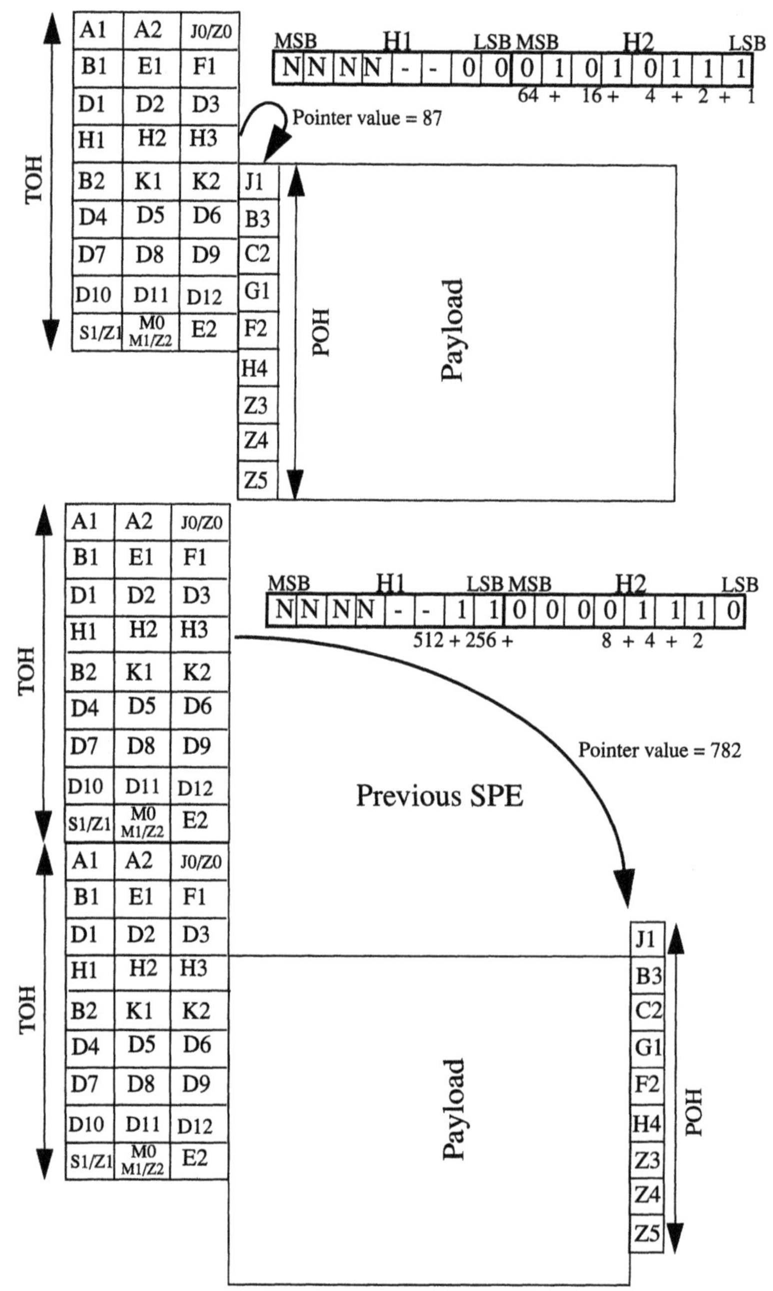

FIGURE 6.4. Example of the STS-1 pointer (cont.).

The SPE is not expected to shift often because everything is supposed to be synchronized. Consecutive pointer operations do not change for at least three frames in which the pointer value remains constant. When an SPE does float, it shifts by only one octet at a time (STS-1).

When the frame rate of the STS SPE is slower than that of the TOH, then the alignment of the SPE is periodically slipped back in time and the pointer is incremented by one. This operation is indicated by inverting five I-bits of the pointer word. A positive stuff octet appears immediately after the H3 octet in the frame containing the inverted I-bits as shown in Figure 6.5. Subsequent pointers contain the new offset.

This operation pushes the start of the next SPE to the right one column. The value of the positive stuff octet (octet after H3) is undefined and is ignored by the receiver.

When the frame rate of the STS SPE is faster than that of the TOH, then the alignment of the SPE is periodically advanced in time and the pointer is decremented by one. This operation is indicated by inverting five D-bits of the pointer word. A negative stuff octet appears in the H3 octet position in the frame containing the inverted D-bits as shown in Figure 6.6. Subsequent pointers contain the new offset.

The negative stuff octet (H3) contains one octet from the payload in the case of STS-1. The H3 octet is really a space holder. Normally this position contains no information. When the adjustment needed to move the SPE ahead in time relative to the STS frame, the SPE fills the H3 octet, with payload in one frame, to move the start of the next SPE to the left one column. Thus, a frame slip is avoided.

The increment or decrement decision is made at the receiver by a match of 8 or more of the 10 I- and D-bits to either the increment or decrement indication.

New Data Flag for STS-1 SPE

New connections, fault recovery, and other events may require a drastic adjustment to the pointer that cannot be handled one octet at a time. The new data flag (NDF) is set when a pointer value must change by more than one, because of a shift in the floating SPE by more than one octet.

The N-bits of the pointer word carry an NDF. During normal operation, four N-bits are set to 0110. The NDF is set by inverting all four N-bits to 1001. The new alignment of the STS SPE is indicated by the pointer value accompanying the set NDF and is effective at that offset. A normal pointer appears in the succeeding frame, with N-bits at 0110 shown in Figure 6.7.

The NDF decision is made at the receiver by a match of 3 or 4 of the N-bits.

The receiver is expected to recover framing within a half millisecond. This is fast enough to be ignored by voice users.

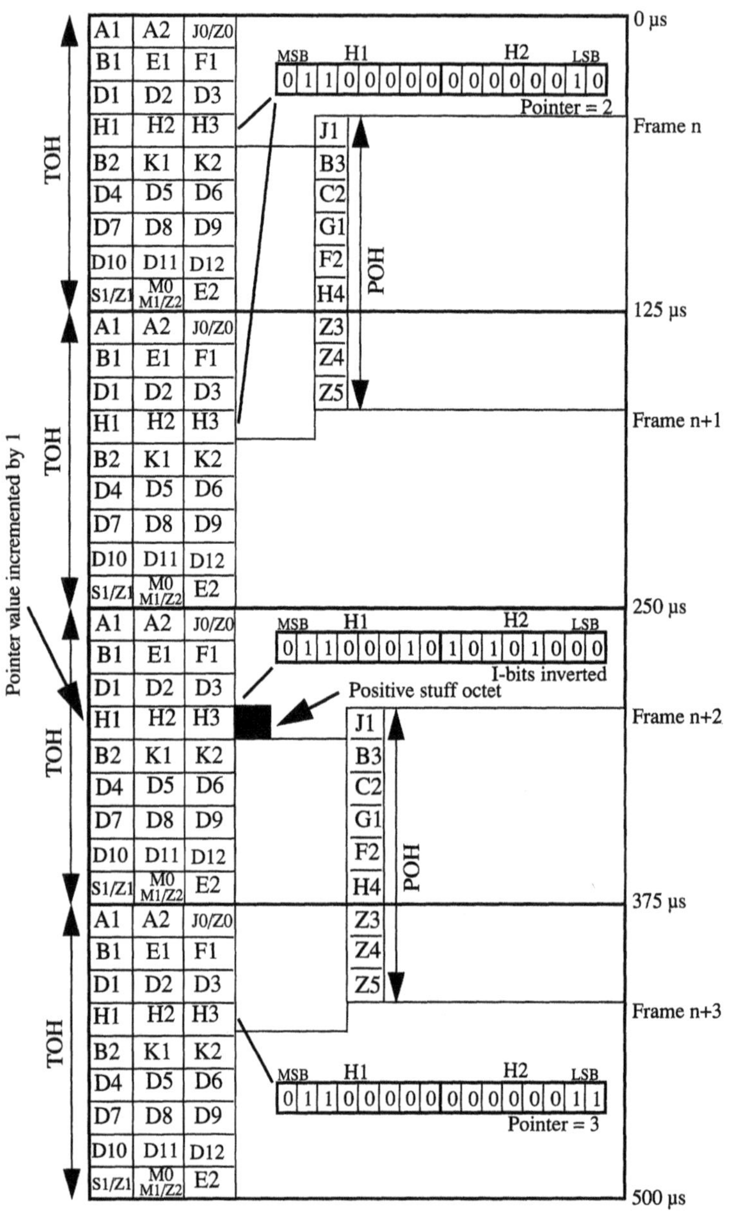

FIGURE 6.5. Positive STS-1 pointer adjustment operation (increment).

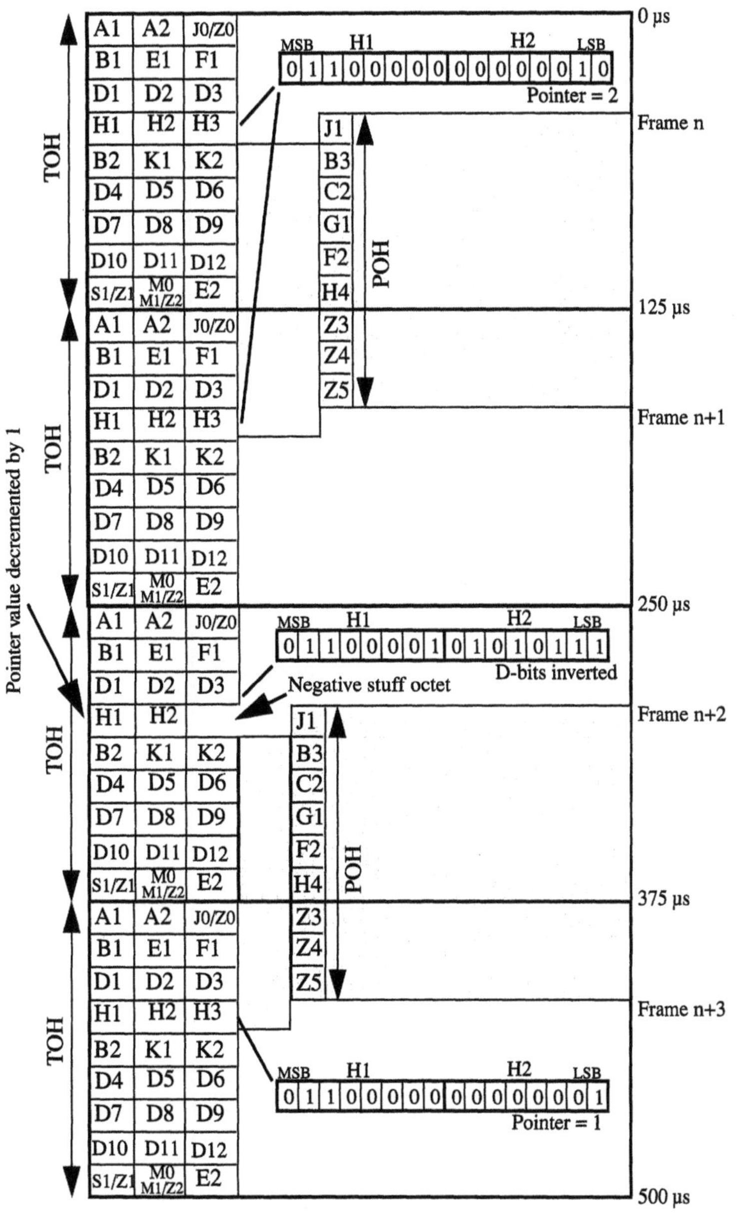

FIGURE 6.6. Negative STS-1 pointer adjustment operation (decrement).

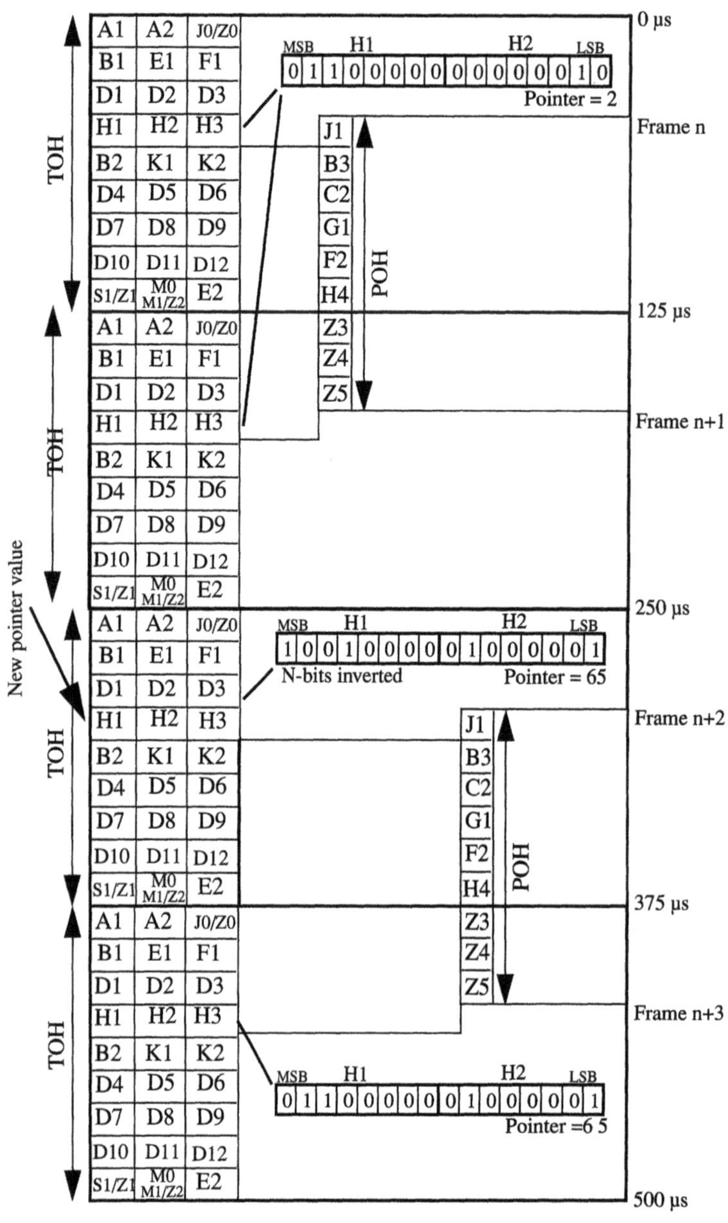

FIGURE 6.7. NDF pointer adjustment operation.

Concatenation Indicator

A concatenation indicator contained in the payload pointers of the second through Nth STS-1s in an STS-Nc are used to show that those STS-1s each contain part of the STS-Nc SPE, not individual STS-1 SPEs.

The first STS-1 within an STS-Nc contains a normal pointer word. All subsequent STS-1s within an STS-Nc have their pointer value set to "1001 xx 11 1111 1111," which is the concatenation indicator.

VT Payload Pointers

The VT payload pointer provides a mechanism that allows the flexible and dynamic alignment of the VT SPE within the VT superframe. Dynamic alignment means that the VT SPE is allowed to float within the VT superframe. Therefore, the pointer is able to accommodate phase differences of the VT superframe and the STS-1 SPE. It also accommodates differences in the frame rates of the VT superframe and the STS-1 SPE.

VT Pointer Value

The VT payload pointer is contained in the V1 and V2 octets. The V1 and V2 octets are viewed as one word just like H1 and H2 octets as shown in Figure 6.8.

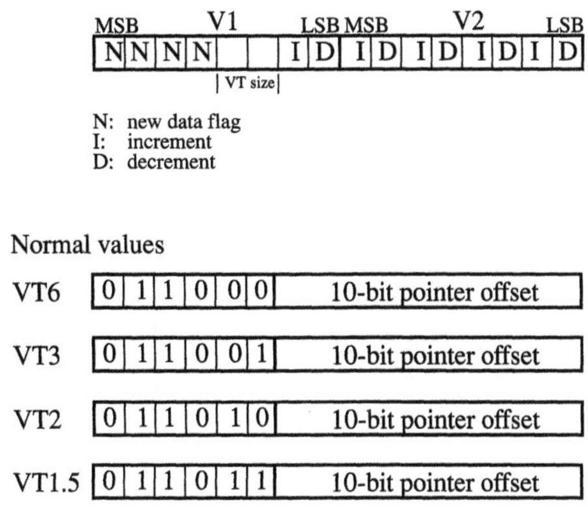

FIGURE 6.8. VT payload pointer (V1, V2).

The four N-bits of the pointer word carry the NDF, and the 10 I-bits and D-bits carry the pointer value. The remaining two bits define the VT sizes. The pointer value is a binary number with a range of 0 to 103 (VT1.5), 0 to 139 (VT2), 0 to 211 (VT3), or 0 to 427 (VT6), that indicates the offset between the pointer word and the first octet of the VT SPE as shown in Figure 6.9.

For example, a pointer value of 0 indicates that the SPE starts in the octet position immediately following the V2 octet.

VT Frequency Justification

When there is a frequency offset between the frame rate of the VT SPE and that of the STS-1 SPE, then the pointer value will be incremented or

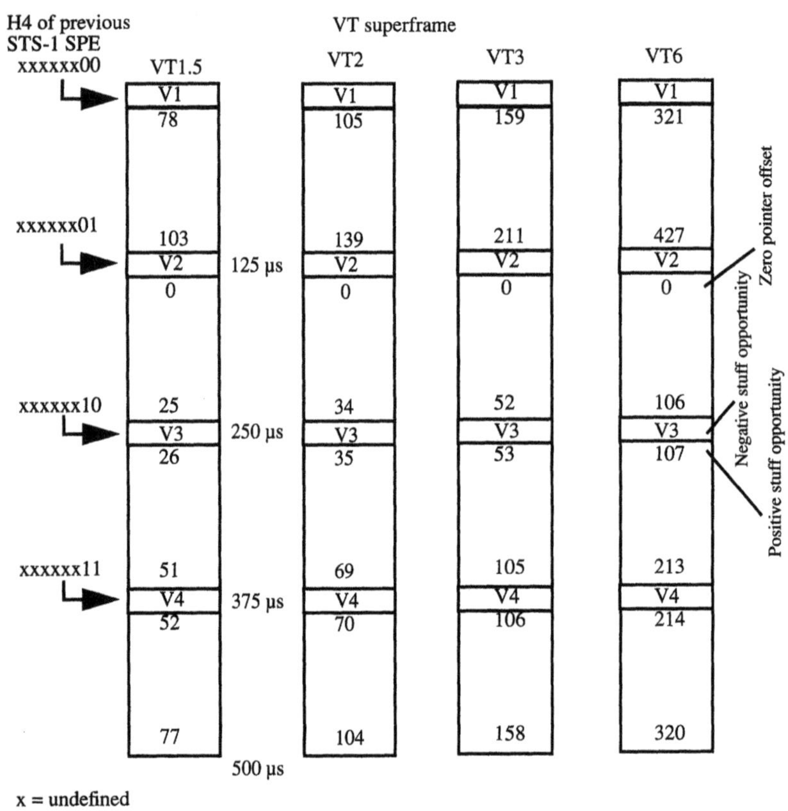

FIGURE 6.9. VT pointer offsets.

decremented, as needed. This is accomplished by the use of a positive or a negative stuff octet.

The VT SPE is not expected to shift often because everything is supposed to be synchronized. Consecutive pointer operations do not change for at least three frames in which the pointer value remains constant. When an VT SPE does float, it shifts by only one octet at a time.

When the frame rate of the VT SPE is slower than that of the STS-1 SPE, then the alignment of the VT SPE is periodically slipped back in time and the pointer is incremented by one. This operation is indicated by inverting five I-bits of the pointer word. A positive stuff octet appears immediately after the V3 octet in the frame containing the inverted I-bits. Subsequent pointers contain the new offset.

This operation pushes the start of the next VT SPE one octet forward. The value of the positive stuff octet (octet after V3) is undefined and is ignored by the receiver.

When the frame rate of the VT SPE is faster than that of the STS-1 SPE, then the alignment of the VT SPE is periodically advanced in time and the pointer is decremented by one. This operation is indicated by inverting five D-bits of the pointer word. A negative stuff octet appears in the V3 octet position in the frame containing the inverted D-bits. Subsequent pointers contain the new offset.

The negative stuff octet (V3) contains one octet from the payload. V3 octet is really a space holder. Normally this position contains no information. When the adjustment needed to move the SPE ahead in time relative to the STS frame, the SPE fills V3 octet, with payload in one frame, to move the start of the next SPE to the left one column. Thus, a frame slip is avoided.

The increment or decrement decision is made at the receiver by a match of 8 or more of the 10 I- and D-bits to either the increment or decrement indication.

NDF for VT SPE

New connections, fault recovery, and other events may require such a drastic adjustment to the pointer they cannot be handled one octet at a time. The NDF is set when a pointer value must change by more than one, because of a shift in the floating VT SPE within STS-1 SPE by more than one octet.

The N-bits of the pointer word carry an NDF. During normal operation, four N-bits are set to 0110. The NDF is set by inverting all four N-bits to 1001. The new alignment of the STS SPE is indicated by the pointer value accompanying the set NDF and is effective at that offset. A normal pointer appears in a succeeding frame, with N-bits at 0110.

The NDF decision is made at the receiver by a match of three or four of the N-bits.

CHAPTER 7

Error Control

Introduction

Any data transmission is subject to error. Because of attenuation, attenuation distortion, delay distortion, noise, and component malfunction, the receiving end may receive data from the transmitting end with some bits altered. Therefore, some form of error control is needed to avoid delivering incorrect data to the user. Error control includes:

- Error detection;
- Error notification;
- Error correction.

Error Detection

A calculation is performed on the data unit to be transmitted, and the result (error detecting code) is then transmitted along with the data unit. On reception, the same calculation is performed on the received data unit and the calculated result is compared with the received error detecting code. If there is a mismatch, the receiver assumes that the received data unit is corrupted.

The following three are the most commonly used error-detecting techniques:

1. Parity;
2. Checksum;
3. CRC (Cyclic Redundancy Check).

Parity

When binary data are transmitted or stored, an extra bit (parity bit) is added for error detection. This is the simplest error detection method. For example, if data are being transmitted in groups of eight bits, a ninth bit is added to each group. When the total number of 1-bits in the block are odd, it is known as odd parity and when the total number of 1-bits in the block are even, it is known as even parity.

Figure 7.1 shows the parity generator and checker block in a system where the transmitter contains the parity generator block and the receiver contains the parity checker block. The transmitted data are also passed through the parity generator block to create the parity bit. The parity bit is then transmitted along with the data to the receiver through the transmission line. The receiver then generates the same parity bit using the received data and compares the generated parity bit with the received parity bit. They should match if there is no error in transmission. Examples of 8-bit data with a parity bit is shown in Table 7.1. Figure 7.2 shows a parity generator circuit.

Checksum

Checksum is the method where all octets of a packet are added together to produce a checksum octet, which is then transmitted at the end of the packet. The receiver also generates its own checksum octet and compares with the received checksum octet. Any mismatch flags an error.

CRC

CRC is one of the most powerful and widely used error-detecting codes today. The theory of CRC is as follows:

- D = Original data block to be transmitted;
- d = Length of the original data block in bits;

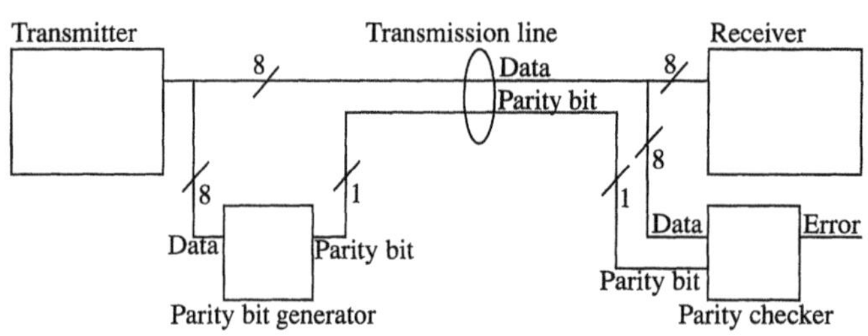

FIGURE 7.1. Parity generator and checker block.

TABLE 7.1. Example of Even and Odd Parity

8 data bits	Parity bit (even parity)	Parity bit (odd parity)
00000000	0	1
00000001	1	0
01001000	0	1
00010101	1	0
10001000	0	1
11001001	0	1

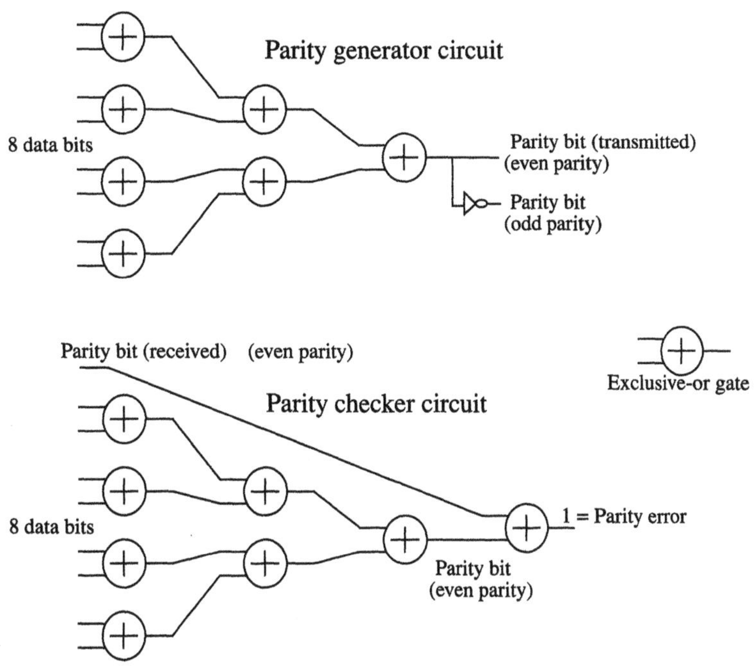

FIGURE 7.2. Example of parity generator and checker.

- *FCS* = Frame check sequence, which is generated by the transmitter;
- *l* = Length of *FCS* in bits, where $l < d$;
- *M* = Transmitted message of $d+l$ bits long, where *M* is the concatenation of *D* and *FCS*;
- *CRC-n* = Predetermined divisor of $l + 1$ bits.

Given *D*, a *d*-bit original block of data. The transmitter generates a l-bit *FCS* so that the resulting (d + l)-bit message *M*, consisting of *D* and *FCS*, is exactly divisible by a predetermined divisor *CRC-n*. The receiver then divides the incoming message *M* by the same predetermined divisor *CRC-n* and, if there is a remainder, it assumes that the original data are corrupted. Therefore

$$M = 2^l D + FCS$$

The original data (*D*) can be represented as a polynomial *X*, where the individual bits of *D* are the coefficients of polynomial *X*. The values of D_l (0 or 1) are the bits

$$D(X) = D_l X^l + D_{l-1} X^{l-1} + \ldots + D_1 X^1 + D_0 X^0$$

of original data *D*.

An example of the procedure is given below:

- Original data *D* = 11001110, where *d* = 8 bits;
- *CRC*-4 = 1001, predetermined divisor, where *n* = 4 bits;
- Therefore, the *FCS* is (*n*-1) bits long, where *l* = 3 bits.

The original data *D* can be written as follows:

$$D(X) = 1X^7 + 1X^6 + 0X^5 + 0X^4 + 1X^3 + 1X^2 + 1X^1 + 0X^0$$

$$= X^7 + X^6 + X^3 + X^2 + X$$

The *CRC*-4 can also be written as follows:

$$CRC4(X) = 1X^3 + 0X^2 + 0X^1 + 1X^0 = X^3 + 1$$

The original data *D* is multiplied by 2^3 to get 11001110000: this product is then divided by the *CRC*-4.

```
                    11010100
CRC-4 ---->  1001 | 11001110000  <---- D (original data)
                    1001                Shifted left 3 times
                    ----
                     1011
                     1001
                     ----
                       1011
                       1001
                       ----
                         1000
                         1001
                         ----
                          100  <---- FCS (remainder)
```

Therefore, the transmitted message is 11001110100.

The receiver receives the message *M* and then divides it by *CRC*-4.

```
                11010100
CRC-4 ──►  1001 | 11001110100  ◄──── M (received message)
                  1001
                   1011
                   1001
                     1011
                     1001
                       1001
                       1001
                        000  ◄──── No remainder
```

Because there is no remainder, the receiver assumes that the original data are received without any error.

The *CRC* process can easily be implemented, as shown in Figure 7.3 consisting of exclusive-or gates and a shift register equal to the length of *FCS*. There are up to *l* exclusive-or gates. The terms $x2$ and $x1$ are not connected because they are missing from the *CRC*-4 (predetermined divisor polynomial). Initially, the shift register is cleared. The original data are then entered, one bit at a time (MSB first). The first

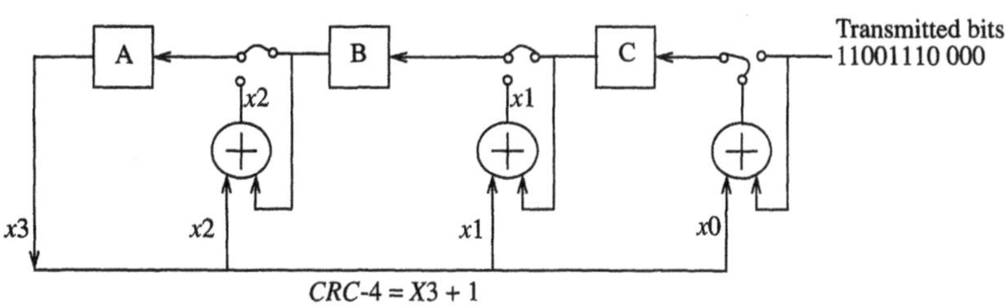

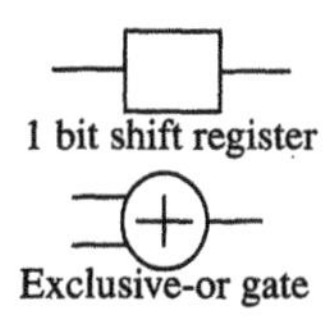

	A B C	Input bits
Initial contents	0 0 0	
Step 1	0 0 1	1
Step 2	0 1 1	1
Step 3	1 1 0	0
Step 4	1 0 1	0
Step 5	0 1 0	1
Step 6	1 0 1	1
Step 7	0 1 0	1
Step 8	1 0 0	0
Step 9	0 0 1	0
Step 10	0 1 0	0
Step 11	1 0 0	0

FSC (remainder)

FIGURE 7.3. CRC circuit.

three bits are shifted left into the registers A, B, and C. Whenever the register A contains a 1, it is added (exclusive-or) to the input of the register C on the next shift. This process is repeated through all the bits of the message including three zeros to accommodate 3-bit *FCS*. At the end of step 11, the shift register contains the FCS, which is transmitted at the end of the message.

At the receiving end, the same circuit is implemented. The received message is then passed through the shift register one bit at a time. At step 9, the input bit will contain 1, which will zero out the register C instead of one. At the end of step 11, registers A, B, and C will contain zeros if there is no error in transmission.

Error Notification

Whenever an error is detected, the error packet is discarded and the control processor of the receiver is notified. A negative acknowledgment and retransmission request may be sent to the transmitting end. If the transmitting end awaits the positive acknowledgment for error-free packets, it will time out and retransmit for those packets that contain errors that are due to not receiving positive acknowledgments.

Error Correction

Error correction is normally done by retransmission. Other methods of correcting errors, such as self-correcting codes, are expensive and require complicated hardware.

CHAPTER 8

Physical Layer

ATM Physical Layer

The ATM Forum has confined itself to working only on the lower two layers of the ATM protocol reference model: the ATM layer and the physical layer. The physical layer is segmented into two sublayers:

1. Physical medium-dependent (PMD) sublayer;
2. Transmission convergence (TC) sublayer.

The physical layer allows ATM interfaces to be built on a wide variety of physical interfaces. The ATM physical layer is shown in Figure 8.1.

Physical Medium-Dependent Sublayer

The PMD sublayer is specific to a particular type of physical medium and deals with such things as bit timing, pulse shape, line coding, jitter, timing recovery, and the physical interfaces (e.g., connectors, coaxial cable, optical fiber). PMD does not include framing or overhead information. This text will cover the following three of the physical mediums for the ATM physical layer: DS-1, DS-3, and SONET.

Transmission Convergence Sublayer

The TC sublayer performs a convergence function and is independent of the physical medium. The convergence function is a process that receives a bitstream from the PMD and extracts cells to pass to the ATM layer. The way these functions are

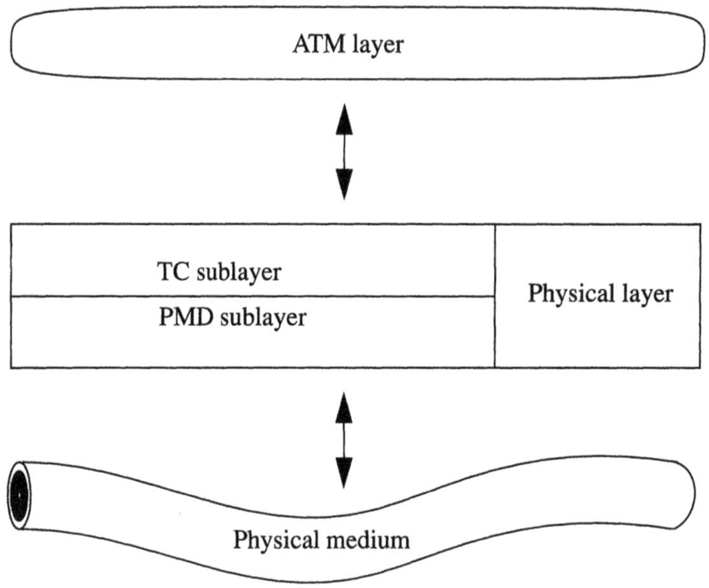

FIGURE 8.1. ATM physical layer.

performed differs depending on the type of physical medium used. Some of the normal functions of the TC sublayer are:

- Cell delineation;
- Cell rate decoupling;
- Header error check (HEC) generation and checking;
- Various operation administration and maintenance (OA&M) functions.

Cell Delineation and Cell Rate Decoupling

Cell delineation is the extraction of cells from the bitstream received from the PMD sublayer. Cell rate decoupling maintains the speed of the ATM layer cellstream to the cell rate of the physical interface. This adaptation is accomplished by the deletion or insertion of idle cells, from or into the ATM cellstream.

The ATM standards also require that the physical layer use the HEC to perform the PMD framing. Therefore, the HEC field is checked by the TC for every incoming cell to ensure that it has properly determined the start and end of a cell by calculating the HEC and checking the result against the HEC of the received cell. The TC will keep searching for the proper cellstream if these do not match for several successive cells.

Header Error Check

The HEC field found in the ATM cell header is used to find the ATM cell boundaries. The HEC is the CRC-8 [$g(x) = x^8+x^2+x+1$] calculation over the first four octets of the ATM cell header. This algorithm determines the valid cell boundary location by searching the 53 possible cell boundaries one at a time. The search state is denoted as the *hunt state*. After a correct HEC is found, it enters the *pre-sync state*. This is the state that searches for the particular cell boundary. If no HEC error is detected, it enters the *sync state*. Synchronization is maintained in this state. When consecutive HEC errors are detected, it returns to the *hunt state*. *Cell delineation state* is shown in Figure 8.2.

Optical Transmission System

There are two ways an optical signal can be transmitted over fiber-optic cable. These are as follows:

1. Single-mode;
2. Multi-mode.

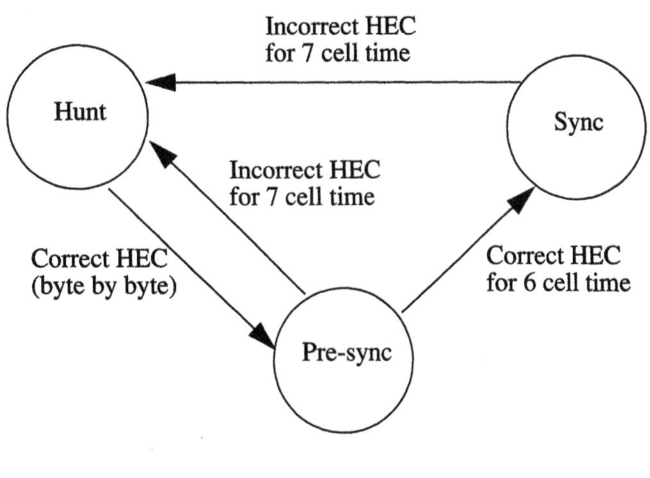

Note: The numbers above are defined by the standards.

FIGURE 8.2. Cell delineation state diagram.

Single-mode uses single-mode fiber cable. The light source is a laser transmitter. The light detector is an avalanche photo diode (APD). The bandwidth is generally higher (~ 100 Gbps) and can service a longer distance (100 km). It is more expensive and normally used by a service provider. On the other hand, *multi-mode* uses multi-mode fiber cable. The light source is a light emitting diode (LED) transmitter. The light detector is a positive intrinsic negative (PIN) or APD. The bandwidth is generally lower (~ 1 Gbps) and can not service longer distances (1 km). It is less expensive and normally used in customer premises equipment. See Figure 8.3 for the single-mode and multi-mode fibers.

DS-1 Physical Layer Interface

The framing of the user data depends on the networking situation and the user's applications. To meet the minimal requirement, every 193rd bit position must follow an SF or ESF pattern. This is true regardless of the information format sent by the user. The information may be ATM packets at the rate of 1.536 Mbps (192 bits/ 125 μs). ATM cells are 53 octets long.

D4 framing (F-bits in the 193rd position) is inserted in the ATM packet. D4 framing is transparent to this type of multiplexer function because the insertion of F-bits at the transmitting end and their removal at the receiving end are done in hardware.

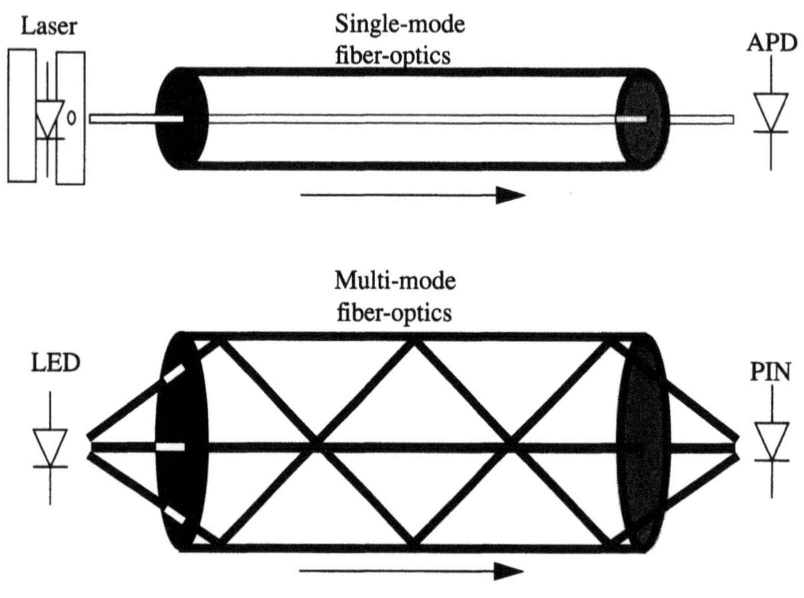

FIGURE 8.3. Single-mode and multi-mode fibers.

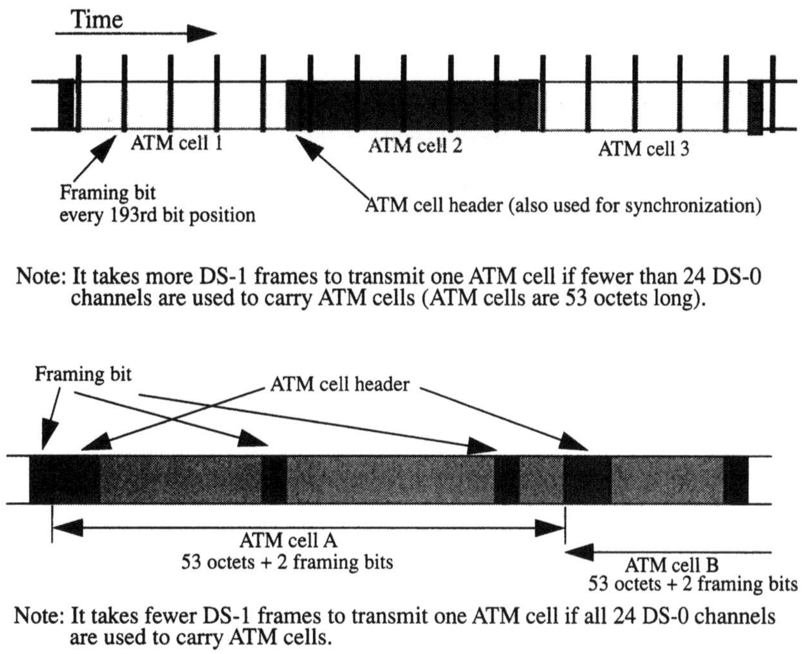

FIGURE 8.4. ATM over DS-1.

Refer to ITU-T G.804 for more information. ATM over DS-1 is shown in Figure 8.4.

DS-3 Physical Layer Interface

Refer to the DS-3 section of Chapter 4 for the DS-3 M-frame format. ATM cell mapping-related information is described in this section.

DS-3 Physical Layer Functions

The DS-3 physical layer functions are grouped into the PMD sublayer and the TC sublayer. These are shown in Table 8.1.

PMD Sublayer Specification

The PMD sublayer deals with DS-3 physical medium criteria (e.g., twisted pair, coaxial cable) defined in ANSI T1.107a and GR-499-CORE.

TABLE 8.1. DS-3 Physical Layer Functions

TC sublayer	HEC sequence generation and verification PLCP framing and cell delineation POH utilization PLCP timing (125 μs clock recovery) Nibble stuffing
PMD sublayer	Bit timing and line coding Physical medium

TC Sublayer Specification

The TC sublayer is independent of the transmission medium characteristics. The main functions of the TC sublayer are to generate and process some overhead octets contained in the DS-3 frame.

ATM Cell Mapping

The physical layer convergence protocol (PLCP) for DS-3 is used to map ATM cells onto existing DS-3 44.736-Mbps communication facilities. The PLCP is then mapped into the DS-3 information payload.

The DS-3 PLCP consists of frame within a standard DS-3 payload and may begin anywhere inside the DS-3 payload.

The DS-3 PLCP frame consists of 12 rows of ATM cells, each preceded by four octets of overhead. Although the DS-3 PLCP is not aligned to the DS-3 framing bits, the octets in the DS-3 PLCP frame are nibble aligned to the DS-3 payload envelope. Nibble stuffing is required after the twelfth cell to frequency-justify the 125 μs PLCP frame. Nibbles begin after the control bits (F, X, P, C or M) of the DS-3 frame. Extraction of ATM cells from the DS-3 operates in the analogous reverse procedure. See Table 8.2 for the DS-3 PLCP Format.

Since the DS-3 frame includes PLCP overhead, the nominal bit rate available for the transport of ATM cells in the DS-3 PLCP is calculated as follows:

Twelve ATM cells/rows x 53 octets/ATM cell x 8 bits/octet = 5,088 bits in 125 μs.

Therefore, the nominal bit rate available for the transport of ATM cells in the DS-3 PLCP = 5,088 bits /125 μs = 40.704 Mbps.

TABLE 8.2. DS-3 PLCP Format

PLCP framing		POI	POH	PLCP payload (ATM cells)	
1	1	1	1	53 Octets	
A1	A2	P11	Z6	ATM cell	
A1	A2	P10	Z5	ATM cell	
A1	A2	P9	Z4	ATM cell	
A1	A2	P8	Z3	ATM cell	
A1	A2	P7	Z2	ATM cell	
A1	A2	P6	Z1	ATM cell	
A1	A2	P5	F1	ATM cell	
A1	A2	P4	B1	ATM cell	
A1	A2	P3	G1	ATM cell	
A1	A2	P2	M2	ATM cell	
A1	A2	P1	M1	ATM cell	
A1	A2	P0	C1	ATM cell	Trailer: 13-14 nibbles

125 μs

12 rows x {(4 octets + 53 octets) x 8 bits/octet} + (13 or 14 nibbles x 4 bits/nibble) =
5,524 bits in 125 μs or
5,528 bits in 125 μs

Therefore, PLCP rate = 5,524 bits / 125 μs = 44.19 Mbps
5,528 bits / 125 μs = 44.22 Mbps

Refer to ITU-T G.804 for more information.

DS-3 PLCP Overhead Octet Description

A1, A2: Frame Alignment Pattern

The A1, A2 are coded with F628H.

The receiver searches for the PLCP frame alignment pattern in the receive stream. When the pattern has been detected for two consecutive rows, along with two valid and sequential path overhead identifier (POI) octets, the UNI declares in-frame. The ATM cell delineation is accomplished by locating the PLCP frame. When errors are detected in both bytes in a single row, or when errors are detected in two consecutive POI bytes, the UNI declares out-of-frame.

P0-P11: Path Overhead Identifier

The twelve unique code values for DS-3 PLCP POI are shown in Table 8.3.

The receiver identifies the location of the PLCP path overhead octets by monitoring the sequence of the POI octets.

Z1-Z6: Growth

These octets are coded with all zeros.

F1: User Channel

These octets are coded with all zeros.

B1: Bit-Interleaved Parity

This octet contains a BIP-8 calculated across the entire PLCP frame (12 rows by 54 octets, excluding the A1, A2, and Pn octets and the trailer nibbles). The B1 value is calculated based on even parity. In the receiving side, the BIP-8 is calculated for the current frame and stored. The B1 octet contained in the subsequent frame is extracted and compared with the calculated value. Differences between the two values provide an indication of the end-to-end bit-error rate.

TABLE 8.3. DS-3 PLCP Path Overhead Identifier (POI)

POI	POI code (hex)	Associated POH
P11	2C	Z6
P10	29	Z5
P9	25	Z4
P8	20	Z3
P7	1C	Z2
P6	19	Z1
P5	15	X
P4	10	B1
P3	0D	G1
P2	08	X
P1	04	X
P0	01	C1

G1: Path Status

The number of B1 errors detected at the near end is conveyed into the first four bit positions, which indicates the PLCP remote error indication (REI). The REI field has nine legal values (0000–1000) indicating between zero and eight B1 errors.

The fifth bit position is used to transmit the PLCP yellow alarm. This bit is set when the receiver detects a loss of PLCP frame (LOF). The last three bit positions are not used and are set to all zeros. The receiver monitors the G1 octet for PLCP status.

M1, M2: Control Information

These octets are coded with all zeros.

C1: Cycle/Stuff Counter

The value of this octet dictates the number of stuff nibbles (13 or 14) at the end of each PLCP frame present. The C1 value is varied in a three-frame cycle where the first frame always contains 13 stuff nibbles, the second frame always contains 14 nibbles, and the third frame contains 13 or 14 nibbles, as shown in Table 8.4.

To lock to an external 8-kHz timing reference, the nibble stuffing in the third frame is varied. For the loop-timed applications, this reference can be the received PLCP frame rate (UNI application). The receiver interprets the trailer nibble length according to the received C1 value.

Nibble stuffing is required because the PLCP bit rate and the DS-3 payload rate do not match without the additional 13 or 14 nibbles added to the PLCP frame. PLCP bit rate is calculated as follows:

12 rows x {(4 octets + 53 octets) x 8 bits/octet} + (13 or 14 nibbles x 4 bits/nibble) = 5,524 bits in 125 μs or 5,528 bits in 125 μs.

Therefore the PLCP bit rate = 5,524 bits/125 μs = 44.192 Mbps

= 5,528 bits/125 μs = 44.224 Mbps.

On the other hand the DS-3 frame bit rate = 44.736 Mbps.

TABLE 8.4. DS-3 PLCP Cycle/Stuff Counter (C1)

C1 (hex)	Frame	Trailer length
FF	1	13
00	2	14
66	3 (no stuff)	13
99	3 (stuff)	14

The DS-3 payload contains 7 subframes/frame x 672 bits/subframe = 4,707 bits.

The DS-3 frame length = 4,760 bits/44.736 Mbps = 106.402 μs, which remains the same for the DS-3 payload.

Therefore, the effective DS-3 payload rate = 4,704 bits/106.402 μs = 44.21 Mbps. This rate matches the PLCP bit rate.

Cell Delineation

The cell delineation function permits the identification of cell boundaries in the payload. The PLCP contains ATM cells in predetermined locations; therefore, the PLCP is sufficient to delineate cells.

HEC Generation and Verification

The HEC code is generated by the entire cell header. The HEC is computed based on a specified polynomial at the transmission end. This field is an 8-bit sequence and is the remainder of the division (modulo 2) by the generator polynomial x^8+x^2+x+1 of the polynomial x^8 multiplied by the content of the header excluding the HEC field. The pattern 01010101 is XORed with the 8-bit remainder before being inserted in the last octet of the ATM cell header. All cells with detected errors in the header are discarded if the error cannot be corrected.

Cell Scrambling and Descrambling

Cell scrambling and descrambling allows the randomization of the cell payload to avoid continuous nonvariable bit patterns, such as "10101010 ..." or "00000000 ..." patterns.

Refer to ITU-T document G.804 for more information relating DS-3 PLCP format.

DS-3 Physical Layer Interface Design Examples

The DS-3 physical layer interface block diagram as shown in Figure 8.5 contains the DS-3 UNI interface block, DS-3 line interface unit, ATM adaptation layer (AAL) processor, and transformers. The DS-3 line interface unit terminates DS-3 line, which deals with bit timing and line characteristics. The AAL processor deals with the ATM adaptation layer, which is described in detail in the AAL chapter. See Figure 8.6 for the DS-3 UNI interface block diagram. It contains the DS-3 encoder and decoder, DS-3 framer, PLCP framer, ATM cell processor, ATM cell FIFO, and the universal test and operations PHY interface for ATM (UTOPIA) interface to the AAL processor.

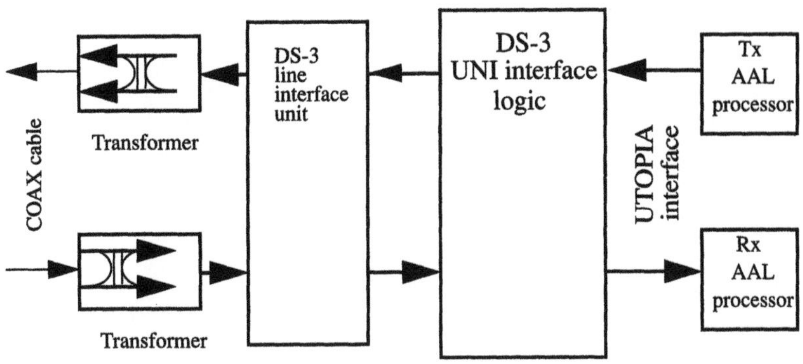

AAL: ATM adaptation layer
UTOPIA: universal test and operations PHY interface for ATM

FIGURE 8.5. DS-3 physical layer interface block diagram.

In the transmit direction, ATM cells are written into the Tx ATM cell FIFO through the UTOPIA interface. ATM cell processor performs functions that are related to the ATM layer, which is discussed later in the ATM layer chapter. The PLCP framer creates PLCP frames by adding PLCP overhead on to 12 ATM cells and passes them to the DS-3 framer. DS-3 frames are created at the DS-3 framer. These DS-3 frames are then passed to the DS-3 line interface unit for transmission. In the receive direction, the DS-3 signal is decoded and passed to the DS-3 framer, where the DS-3 payload (PLCP frame) is extracted. The PLCP processor terminates PLCP frames and passes ATM cells to the ATM processor to terminate ATM cells. These cells are then passed to the AAL processor through the UTOPIA interface.

SONET Physical Layer Interface

The transmission system is based on the SONET standards. The SONET format is developed to define a synchronous optical hierarchy that is flexible enough to carry many different types of payload. A byte-interleaved multiplexing scheme is adopted in SONET with a basic rate of 51.84 Mbps. Refer to Chapter 5 for more information on SONET.

SONET/SDH is an international standard, and the SONET-hierarchy-based interfaces will be the means of having interoperability for both the private and public UNI. These standards provide, through a framing structure, the payload envelope necessary for the transport of ATM cells. SONET also carries operation and maintenance (OAM) information in its overhead octets.

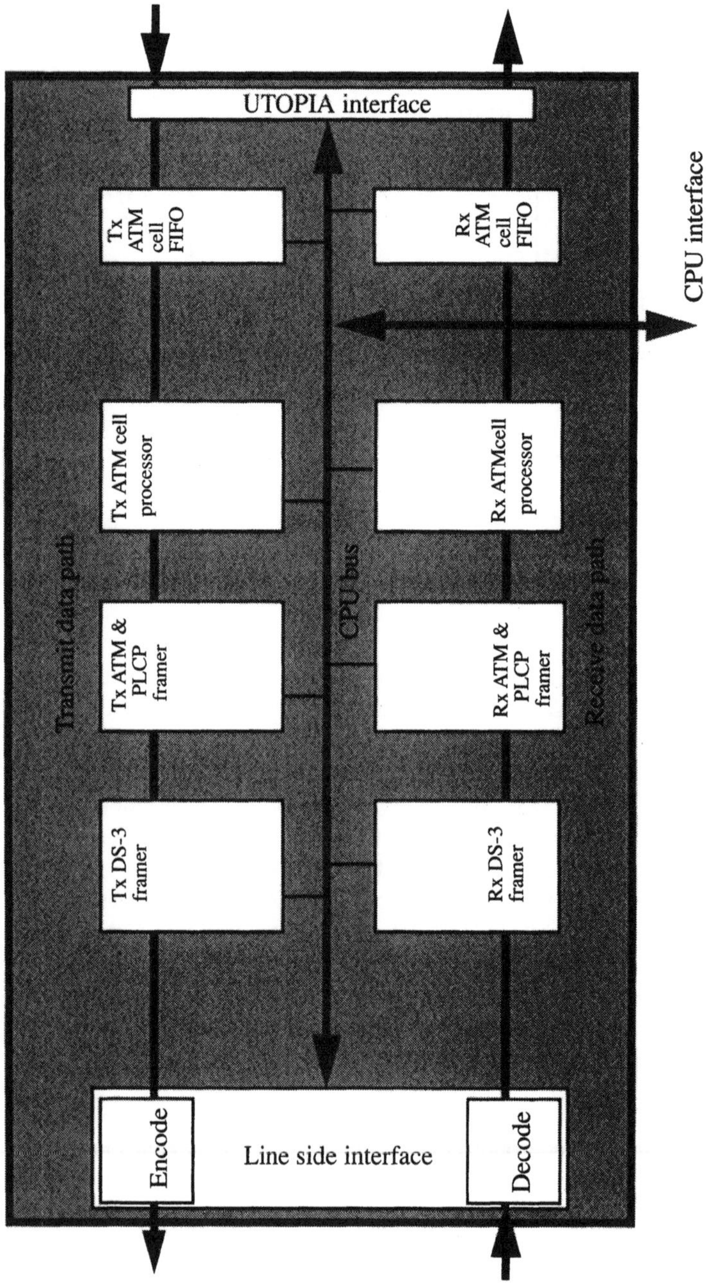

FIGURE 8.6. DS-3 UNI interface block diagram.

SONET Physical Layer Functions

The functions of the physical layer (U-plane) are grouped into the PMD sublayer and the TC sublayer.

See Figure 8.7 for the SONET physical layer functions (U-Plane).

PMD Sublayer Specification

The PMD sublayer deals with SONET physical medium criteria (e.g., OC-3 SMF, OC-3 MMF) defined in ANSI T1.105 and GR-253 CORE.

TC Sublayer Specification

The TC sublayer is independent of the transmission medium characteristics. The main functions of the TC sublayer are to generate and process some overhead octets contained in the SONET STS.

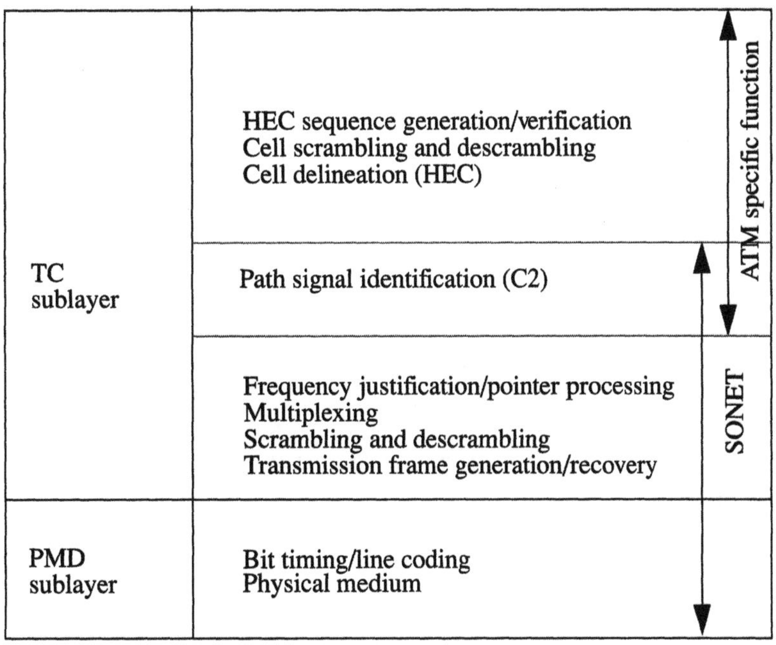

FIGURE 8.7. SONET physical layer functions (U-plane).

ATM Cell Mapping

The SONET STS payload capacity (e.g., synchronous payload envelope) is used to map ATM cells. The entire STS payload capacity is filled with ATM cells. It is indicated through the STS path signal label (C2) byte in the STS POH.

A cell may cross an SPE boundary because the STS payload capacity is not an integer multiple of the ATM cell length.

STS Path Signal Label (C2)

The C2 octet of the STS POH indicates the content of the STS SPE, including the status of the mapped payloads. This octet is coded with 13 hex to indicate that the payload is filled with ATM cells.

ATM Cell Offset (H4)

The H4 octet of the STS POH points to the first ATM cell (53-octet-long) from the J1 octet; later cells begin at 53-octet intervals. This octet indicates the offset in octets to the next ATM cell boundary in the transmit stream. This octet can be used to delineate cell boundaries in the receive stream. Cell delineation techniques that use the HEC octet are preferred. This pointer is generally redundant as the HEC mechanism for cell delineation provides the same function with better performance. H4 octet ranges from 0 to 52.

Cell Delineation

The cell delineation function permits the identification of cell boundaries in the payload. The HEC field in the cell header is used for the cell delineation algorithm. Return to Figure 8.2 for a cell delineation state diagram.

HEC Generation and Verification

The HEC code is generated by the entire header (including the HEC octet) and is contained in the fifth octet of the ATM cell. The HEC sequence code is capable of the following two functions:

- Single-bit error correction;
- Multiple-bit error detection.

The HEC is computed based on a specified polynomial at the transmission end. At the receiver, the errors in HEC can be corrected in the correction mode and detected in the detection mode. In correction mode, only a single-bit error can be corrected, while detection mode provides for multiple-bit error detection. All cells with detected errors in the header are discarded if the error cannot be corrected.

HEC generator polynomial:

$$g(x) = x^8 + x^2 + x + 1$$

Cell Scrambling and Descrambling

Cell scrambling and descrambling allows the randomization of the cell payload to avoid continuous nonvariable bit patterns. ATM cell payload is scrambled with the self-synchronizing generator polynomial $x^{43}+1$.

ATM Mapping for STS-1

One of the physical layers for both the public and private UNI for ATM is the 51.84-Mbps STS-1 frame. ATM cells consists of a 5-octet cell header and a 48-octet payload. ATM cells are mapped into the STS-1 payload by aligning the octet structure of every cell with the octet structure of the STS-1 SPE. The entire STS-1 payload is filled with ATM cells, yielding a transfer capacity for ATM cells of 48.384 Mbps.

Since the STS-1 payload is not an integer multiple of the 53-octet ATM cell length, some cells cross the STS-1 SPE boundary, as shown in Figure 8.8. The ATM cells are mapped onto the payload of the STS-1 SPE. The value of the H4 octet signifies

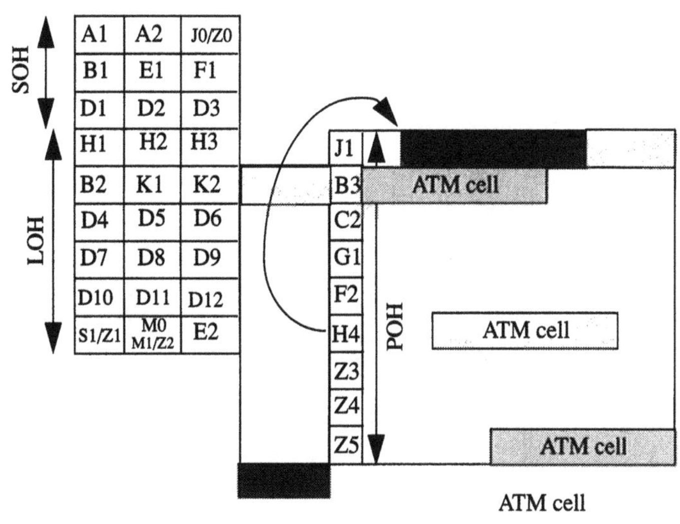

FIGURE 8.8. Mapping for ATM payload in STS-1.

the position of the first ATM cell in the STS-1 SPE, which is the number of octets from the J1 octet.

ATM Mapping for STS-3c

One of the physical layers for both the public and private UNI for ATM is the 155.52 Mbps STS-3c frame.The SONET STS-3c payload capacity (synchronous payload envelope) is used to map ATM cells. The entire STS-3c payload capacity is filled with ATM cells, yielding a transfer capacity for ATM cells of 149.76 Mbps. It is indicated through the STS path signal label (C2) octet in the STS-3c POH. A cell may cross an SPE boundary because the STS-3c payload capacity is not an integer multiple of the cell length. See Figure 8.9.

Physical Layer OAM Specification (M-Plane)

The SONET OAM functions residing in the physical layer management are grouped into three categories:

1. Performance monitoring;
2. Fault management;
3. Facility testing.

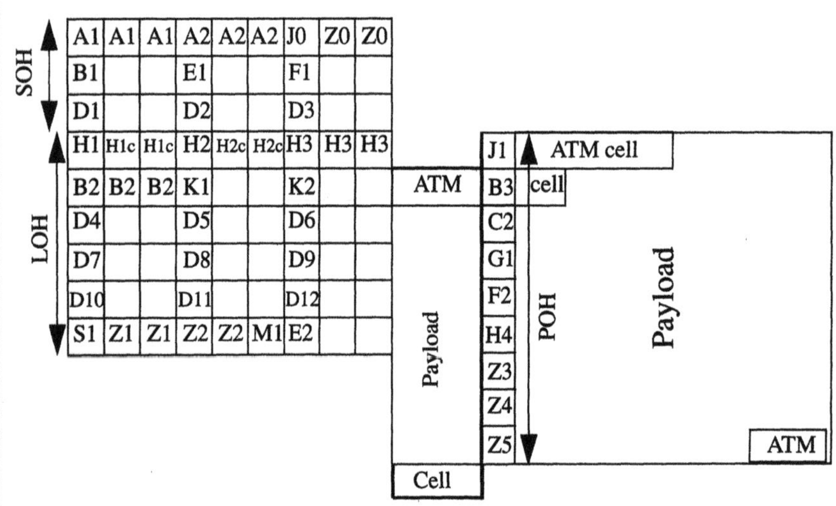

FIGURE 8.9. Mapping for ATM payload in STS-3c.

The OAM flows exchange operations information between nodes in the network access (including customer premises nodes). The F1, F2, and F3 flows are defined at the physical layer. The exchange of information is performed by way of well-defined overhead octets within the SONET framing structure. The F4, F5 level carries the ATM layer information flows over OAM cells. The ITU-T I.610 gives a detailed explanation of OAM layers and information flows. Figure 8.10 shows the SONET physical layer OAM flow at the UNI.

B-ISDN–Independent OAM Functions

The OAM function at the UNI are performed by SONET section, line, and path terminating equipment. These are B-ISDN–independent OAM functions.

Performance Monitoring

Performance monitoring is performed to gather information about the network element behavior in order to evaluate and report on network performance. The performance objectives are divided into the following three major categories:

- Delay objectives;
- Accuracy objectives;
- Availability objectives.

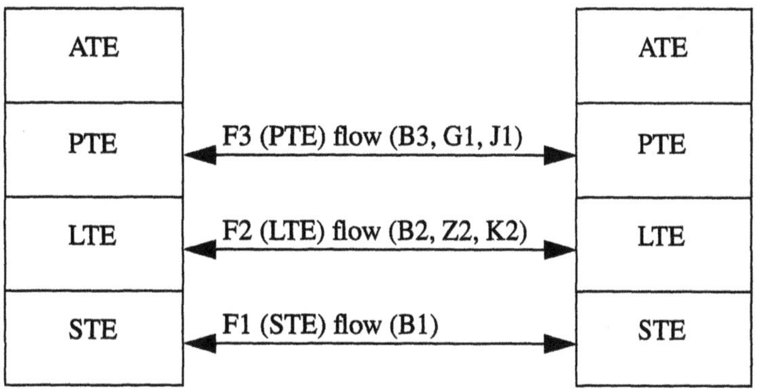

ATE: ATM terminal equipment
PTE: SONET path terminal equipment
LTE: SONET line terminal equipment
STE: SONET section terminal equipment

FIGURE 8.10. SONET physical layer OAM flow at the UNI.

The coding violations are detected by using the B1 octet at the section level; B2, Z2 octets at the line level; and B3 octets, G1(1-4) bits at the path level. Monitoring is performed across the UNI by calculating section BIP-8, line BIP-8, and path BIP-8 of the incoming signal, and then the values obtained are compared with the one encoded in the proper bytes by the transmitting end. The number of BIP errors detected by the line or path terminating equipment is conveyed back to the upstream equipment by using the line and path remote error indication (REI-P) signals.

Fault Management

The fault management functions detect, isolate, and correct failure conditions in the network. Fault management actions at the UNI can be triggered by the following conditions:

- Incoming signal failures;
- Equipment failures;
- Detection or removal of alarm indication signal (AIS);
- Detection or removal of remote defect indicator (RDI) signal.

The failures detected on the incoming signal can be listed as:

- Loss of signal (LOS);
- Loss of frame (LOF);
- Loss of pointer (LOP);
- Signal label mismatch.

Facility Testing (Path Trace)

The verification of the connection continuity in PTE is performed by repetitively sending the appropriate 64-octet code in the J1 POH octet. More detailed information can be obtained from ANSI T1.105 and GR-253-CORE.

B-ISDN–Specific OAM Functions

The following two OAM functions at the UNI are B-ISDN specific:

1. Line-error monitoring;
2. Loss of cell delineation.

Line Error Monitoring

This is a modified or extended SONET function for the UNI. The transmission performance on the outgoing link of the far end (originating) LTE is alerted by the receiving LTE. The REI count is conveyed in the Z2 octet of the third STS-1.

Loss of Cell Delineation

The SONET Path RDI alarm is used to alert the upstream SONET PTE when a loss-of-cell-delineation event occurs downstream.

The physical layer OAM functions are summarized in Figure 8.11.

SONET Physical Layer Interface Design Examples

The SONET physical layer interface block consists of electrical to optical (E/O), optical to electrical (O/E), clock and data recovery, UNI interface logic, and the UTOPIA interface to the ATM AAL processor. Single-mode or multi-mode fiber interfaces directly to the E/O and O/E devices. The clock and data recovery device is necessary to recover the 155.52-Mbps clock signal and data from the incoming signal. This block contains a phase-locked loop. UNI interface logic contains all the circuits to process the STS-3c SONET frame and has a UTOPIA interface to communicate with the ATM adaptation layer. See Figure 8.12 for a SONET physical layer interface block diagram.

Functions	SONET overhead octets	Description
Performance monitoring	HEC error corrected/ uncorrected	Cell HEC error monitoring
	B2, 3rd Z2 byte (2-8)	Line error monitoring
	B3, G1 (1-4)	Path error monitoring
	B1	Section error monitoring
Fault management	H1, H2, H3	STS path AIS
	G1(5)	STS path RDI
	G1 (5)	Loss of cell delineation, path RDI (note)
	K2 (6-8)	AIS, RDI
Facility testing	J1	Path trace, connectivity verification

B-ISDN-specific functions

SONET-specific functions

Note: Loss of cell delineation generates STS path RDI.

FIGURE 8.11. SONET physical layer OAM functions.

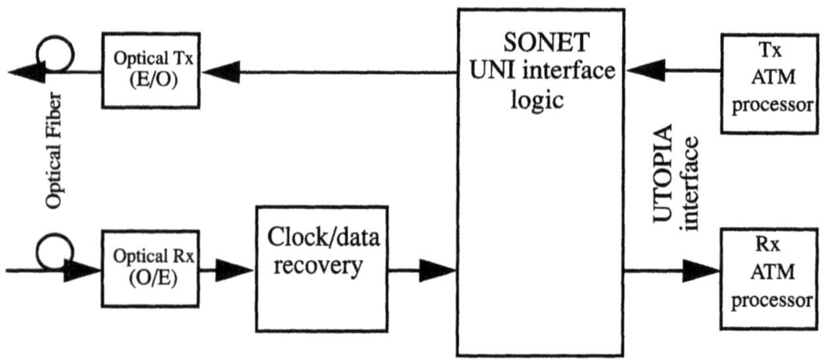

FIGURE 8.12. SONET physical layer interface block diagram.

The UNI interface logic as shown in Figure 8.13 contains the following functional units for both receive and transmit sections:

- Serial/parallel, parallel/serial converter logic;
- Rx/Tx SOH processor;
- Rx/Tx LOH processor;
- Rx/Tx POH processor;
- Rx/Tx ATM cell processor;
- Rx/Tx ATM cell FIFO;
- UTOPIA interface to the AAL processor.

In the transmit direction, ATM cells are written into the Tx ATM cell FIFO through the UTOPIA interface. The ATM cell processor performs functions that are related to the ATM layer, which is discussed later in Chapter 9 on ATM layer. The SONET path, line, and section overheads are added to the payload at the path, line, and section O/H processors, respectively. The octet-aligned SONET frame is then passed to the parallel/serial converter logic, where serial electrical pulses are created, which are then converted to optical pulses.

In the receive direction, the serial/parallel converter logic receives electrical pulses that were originally converted from optical pulses. The serial signal is then converted to an octet-aligned parallel signal. The section, line, and path O/H processors terminate section, line, and path overhead octets, respectively. The ATM cells (SONET payload) are then passed to the ATM processor to terminate ATM cells. These cells are then passed to the AAL processor through the UTOPIA interface. The UTOPIA interface is there to communicate with the ATM adaptation layer.

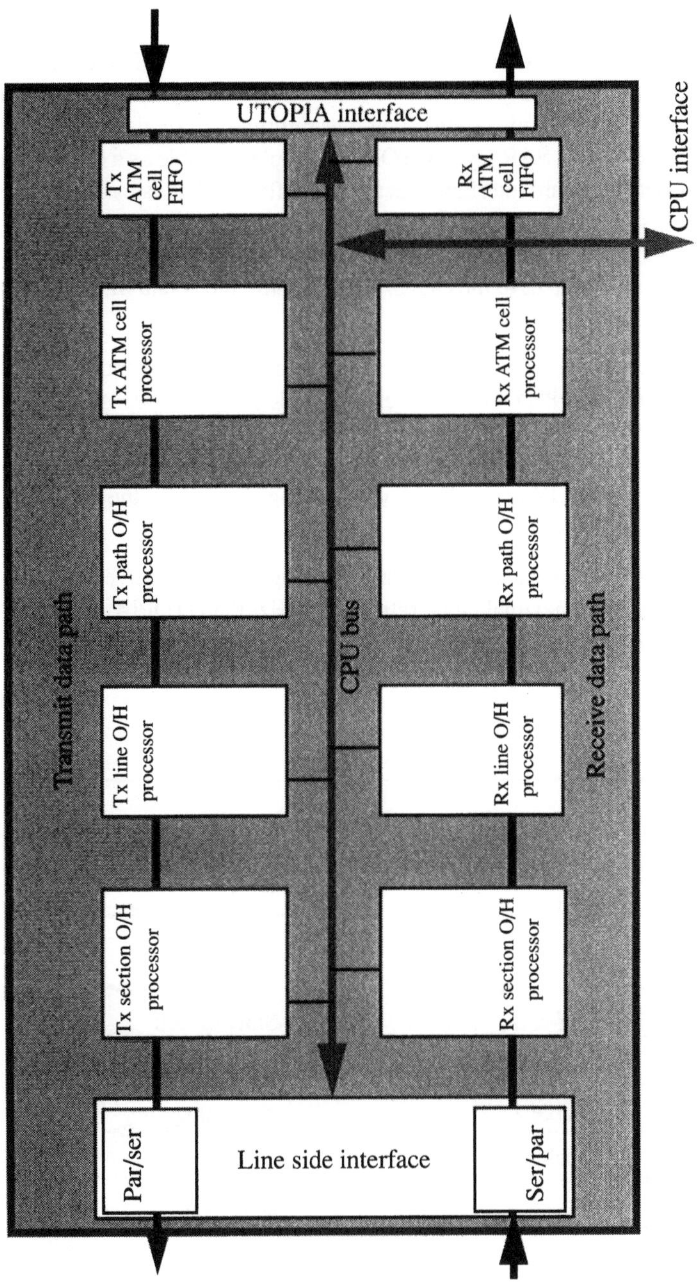

FIGURE 8.13. SONET UNI interface block diagram.

UTOPIA

UTOPIA is a common, standard interface between the segmentation and reassembly (SAR) and physical layers of ATM subsystems. The ATM Forum defined this interface to promote interoperability between various vendors' physical layer (PHY) and SAR functions of the ATM subsystems. It allows multiple physical layers that are supportable by a common SAR layer. UTOPIA is implementable as a chip-internal, chip-to-chip, board-to-board interface. Because of different data rates of the various ATM PHY layers, it is necessary to provide rate-matching buffers (e.g., FIFO's) between SAR and PHY. This interface supports a streaming mode of operation, where both SAR and PHY have the capability to throttle the actual transfer rate.

This UTOPIA applies to the data path interface between the ATM layer (SAR function) and the PHY components of an ATM subsystem. For a multivendor PHY solution, it allows a common PHY interface across a wide range of speeds and media types. This interface addresses many issues, such as an 8-bit or 16-bit data path interface, octet- or cell-level handshaking, and attachment to a single PHY to an ATM layer or multiple PHY to an ATM layer. See Figure 8.14 for the UTOPIA interface to an ATM layer.

Transmit and receive transfers are synchronized using their respective clocks. With an 8-bit data path and a maximal clock rate of 20 MHz, the UTOPIA interface supports the following rates:

- 44.736 Mbps (DS-3);
- 51.84 Mbps (STS-1);

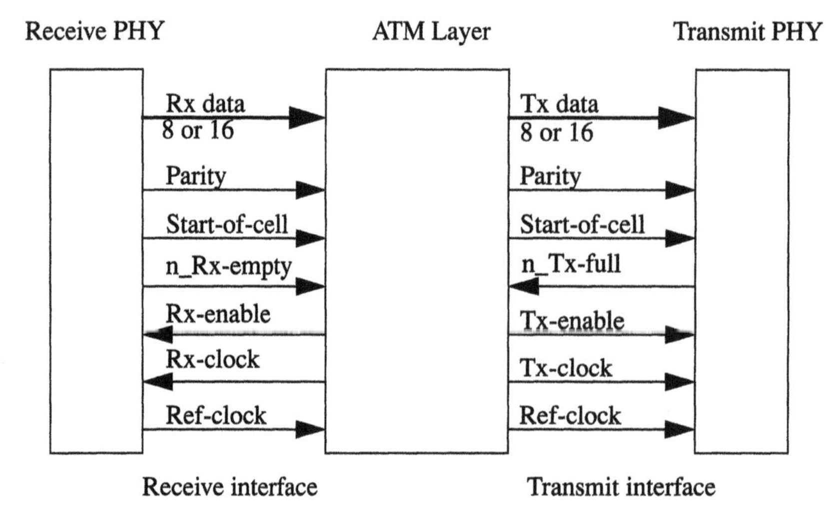

FIGURE 8.14. UTOPIA interface to an ATM layer.

- 155.52 Mbps (STS-3);
- 622 Mbps (STS-12), with a 16-bit data path and a 40-MHz clock.

The transfer of data is synchronized at the cell level using a start of cell (SOC) signal, which is asserted when the data are at the first octet of the ATM cell. The transfer of data at the octet level is by way of a separate transmit and receive clock. Both SAR and PHY throttle the transfer rate by using FIFO and flow control signals.

The receive interface transfers data when the SAR requests it, by asserting an enable signal.

Cell Processing

The transfer of 53 octet cells across the 8-bit data path and 16-bit data path are shown in Figure 8.15.

Transmit Interface Timing

Transmit interface timing is shown in Figure 8.16. TxClk is the data transfer/synchronization clock provided by the SAR to the PHY layer for synchronizing transfers on TxData. TxData is driven from the SAR to PHY. TxSOC is an active high signal asserted by SAR when TxData contains the first octet of the ATM cell. TxEnb is the active low signal asserted by SAR during cycles when TxData contains valid ATM cell data. TxFull is an active low signal from PHY to SAR, asserted by PHY at least four cycles before it is no longer able to accept transmit ATM data.

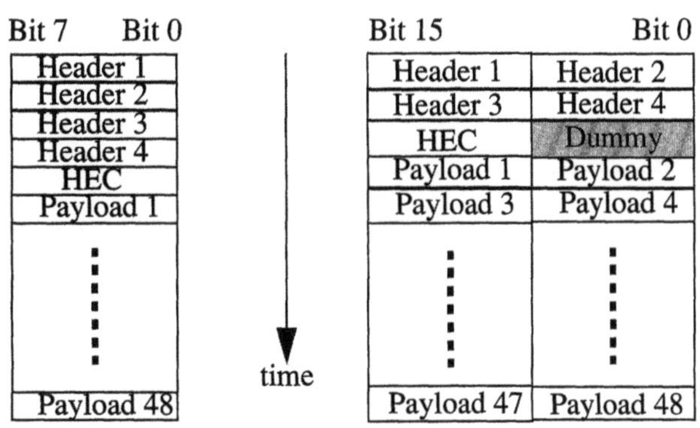

FIGURE 8.15. ATM cell format over UTOPIA interface.

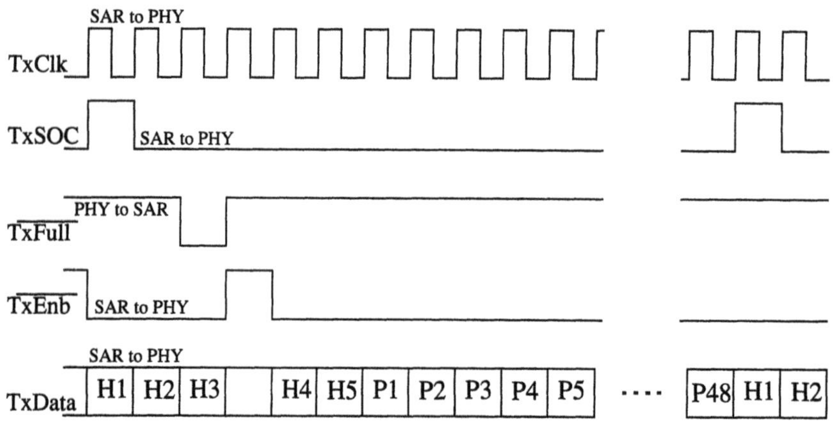

FIGURE 8.16. Transmit interface timing.

Receive Interface Timing

Receive interface timing is shown in Figure 8.17. RxClk is the data transfer/synchronization clock provided by the SAR to the PHY layer for synchronizing transfers on RxData. RxData is driven from the PHY to SAR. RxSOC is an active high signal asserted by PHY when RxData contains the first octet of the ATM cell. RxEnb is the active low signal asserted by SAR to indicate that RxData will be sampled at the start of next cycle. RxEmpty is an active low signal asserted by PHY to indicate that the current cycle contains no valid ATM data.

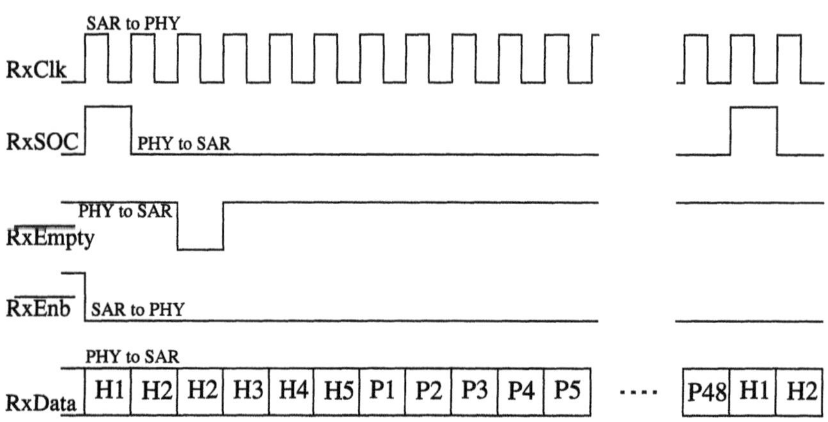

FIGURE 8.17. Receive interface timing.

CHAPTER 9

ATM Layer

ATM Layer Services

The ATM layer deals with the transfer of fixed-size ATM layer protocol data units (ATM-PDU). These ATM-PDUs are called ATM cells. This layer communicates with the AAL layers and the physical layers. This transfer occurs on a preestablished or switched ATM connection according to some traffic contract. The traffic contract is discussed later in this book. The short- and fixed-length ATM cells are the fundamental unit of ATM transmission. Flexibility in bandwidth use is achieved due to short cells. This flexibility in turn provides the basic framework for B-ISDN to support a wide range of services.

The ATM UNI supports two levels of virtual connections:

1. Point-to-point or point-to-multipoint Virtual Channel Connection;
2. Point-to-point or point-to-multipoint Virtual Path Connection.

A point-to-point or point-to-multipoint Virtual Channel Connection (VCC)

It consists of a single connection established between two ATM VCC endpoints.

A point-to-point or point-to-multipoint Virtual Path Connection (VPC)

It consists of a single connection established between two ATM VPC endpoints.

A VPC consists of multiple VCCs carried transparently between two ATM endpoints. VPC and VCC are shown in Figure 9.1.

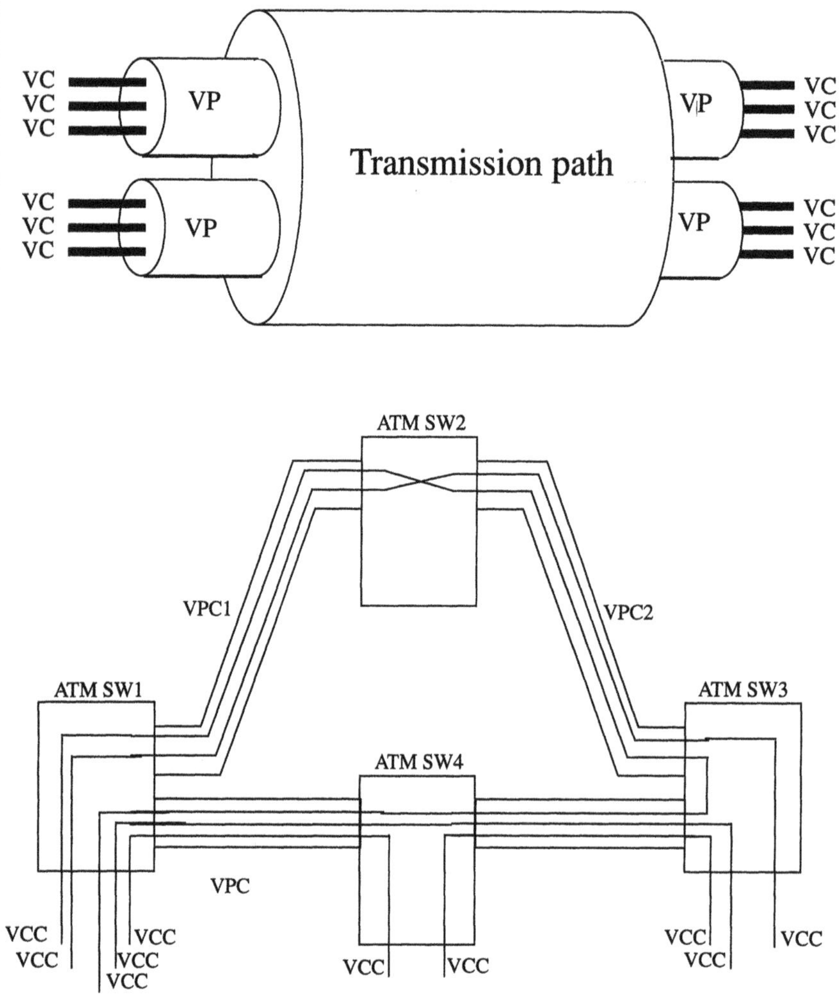

FIGURE 9.1. VPC and VCC.

ATM Cell Structure and Encoding

The same structure of each ATM cell is true within an ATM network (i.e., on an ATM interface or link). The structure of an ATM cell within an ATM switch, however, may vary according to the implementation. The ATM cell header structure is slightly different between a UNI and an NNI interface. See Figure 9.2 and Figure 9.3 for ATM cell (UNI) and ATM cell (NNI) respectively. For the NNI interface the GFC field is removed, and the length of the VPI field is increased.

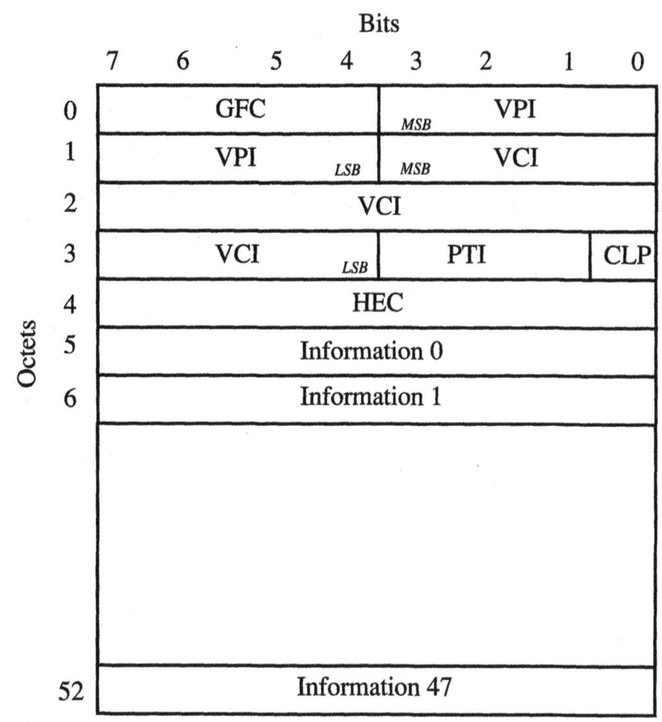

FIGURE 9.2. ATM cell (UNI).

Refer to Bellcore documents GR-1113-CORE, GR-1115-CORE, and ATM Forum document AF-BICI-0013.003 for more information.

ATM Cell Header Fields

The structure of the ATM cell (UNI) header is shown in Figure 9.2. It contains the following fields:

Generic Flow Control (GFC)

The 4-bit GFC field is defined only across the UNI. This field provides a standardized local function that is to control traffic flow on the customer site. The value encoded in the GFC is overwritten by the ATM switches. This field is normally encoded with zeros. Across the NNI, the GFC bits are used for the expanded VPI field.

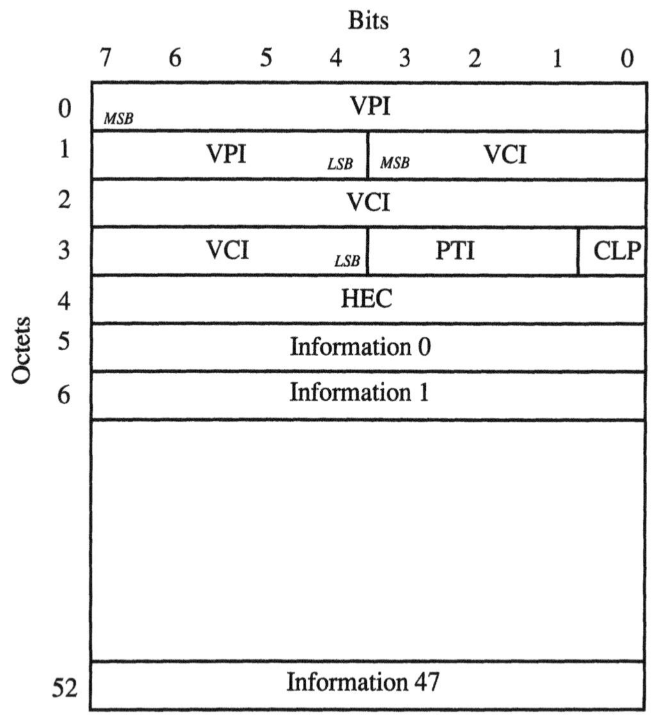

FIGURE 9.3. ATM cell (NNI).

Virtual Path Identifier (VPI)

The VPI field is 8 bits wide across the UNI and 12 bits wide across the NNI. This field is used to identify virtual paths. The default value of the VPI is all zeros. This field is set to zeros for an idle cell.

Virtual Circuit Identifier (VCI)

The VCI field is 16 bits wide across the UNI and NNI. This field is used to identify virtual circuits across either a UNI or an NNI. The default value of the VCI is all zeros. This field is set to zeros for an idle cell.

The entire VPI/VCI space is very large. Therefore, it is not required to support the entire VPI/VCI space. The actual number of routing bits in the VPI and VCI subfields use for routing is negotiated between the user and the network (e.g., on a subscription basis). To make implementations easier and cheaper, the bits within the VPI and VCI fields used for routing are allocated using the following rules:

- The allocated bits of the VPI subfield are contiguous and right justified;
- The allocated bits of the VCI subfield are contiguous and right justified.

Payload Type (PTI)

The PTI field is 3 bits wide. This field is used to identify the payload type carried in the ATM cell. The MSB bit of the PTI field indicates whether it is a user cell or an OAM cell. Within the user cell, the middle PTI bit indicates whether the cell experienced any congestion or not. The LSB bit is used to differentiate between two different types of information payload: type 0 if the bit is 0, type 1 if the bit is 1. Within the OAM cell, the LSB bit is used to differentiate between two different types of OAM flows: segment if the bit is 0; end-to-end if the bit is 1.

Cell Loss Priority (CLP)

The l-bit CLP field allows the user or the network to optionally indicate the cell loss priority. The AAL layer sets the CLP bit. This bit is used by the network to indicate the relative importance of cells. A CLP bit set to 0 indicates a higher priority cell and a CLP bit set to 1 indicates a lower priority cell and can be discarded in the event a switch experiences congestion.

Header Error Check (HEC)

The HEC field is 8 bits wide and contains an 8-bit CRC computed over all fields in the ATM cell header. This field is used by the physical layer for detection and correction of bit errors in the ATM cell header. It is also used for cell delineation.

ATM Cell Payload Size

An ATM cell consists of a 5-octet header and a 48-octet information payload field. The ATM cell length is kept short to provide a great deal of flexibility in terms of bandwidth use. Therefore, it can support a wide range of services that emerging applications require.

The synchronous traffic, such as voice, cannot tolerate large delays. Only a small number of samples can be gathered at a source before these samples have to be transmitted. Therefore, the ATM cells need to be small in size. The voice quality is poor when voice is packetized, such as voice over internet protocol (VoIP) and voice over frame relay (VoFR) due to compression and delay. If ATM cells were very large, most of the cells would not have enough time to be filled before they had to be transmitted, making them very inefficient.

The 48-octet payload format was chosen by the ITU-T as a compromise between the United States, which wanted 64-octet payloads, and Europe and Japan, which wanted 32-octet payloads.

ATM Cell Transmission Order

The octets within the ATM cells are transmitted in increasing order, starting with the first octet of the header, and bits within octets are transmitted in decreasing order (i.e., most significant bit first), starting with bit 8. See Figure 9.4.

ATM Layer Functions Involved at the UNI (U-Plane)

At the UNI, the ATM layer supports a variety of functions, which are shown in Table 9.1.

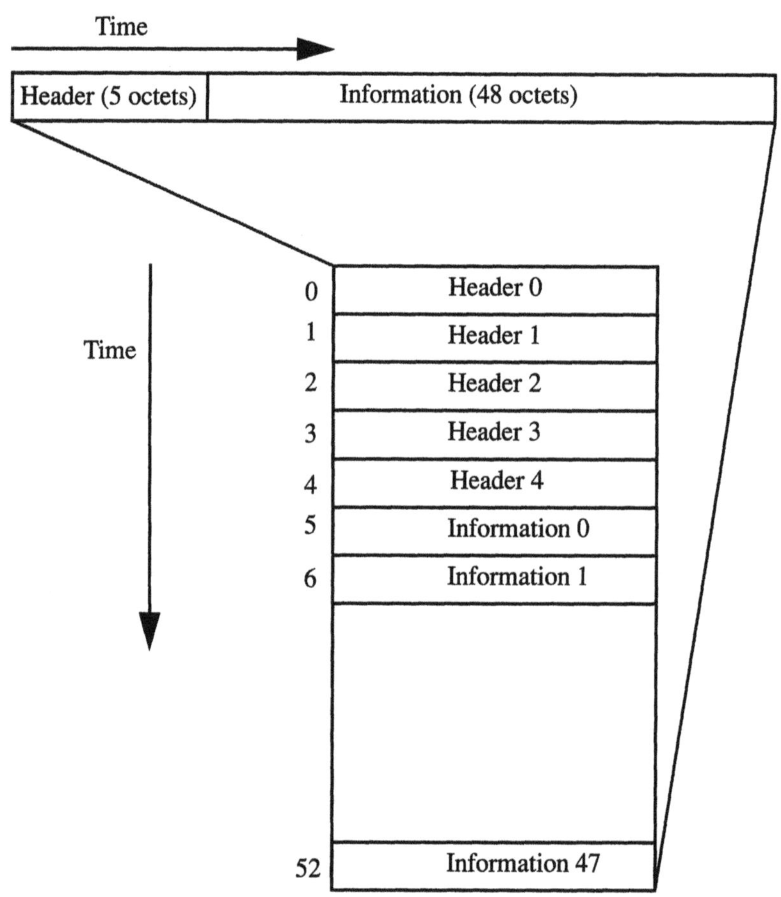

FIGURE 9.4. ATM cell transmission order.

TABLE 9.1. ATM Layer Functions at the UNI (U-Plane)

Functions	Parameters
Multiplexing among different ATM connections	VPI/VCI
Cell rate decoupling (unassigned and idle cells)	Pre-assigned header field values
Cell discrimination based on predefined header field values	Pre-assigned header field values
Payload-type discrimination	PTI field
Loss priority indication and selective cell discarding	CLP field, network congestion state
Traffic shaping	Traffic descriptor

Multiplexing Among Different ATM Connections

This function allows the ATM connections to multiplex different quality of service (QoS)–class traffic. These QoS classes are:

- Class A: circuit emulation, constant-bit rate video;
- Class B: variable-bit rate audio and video;
- Class C: connection-oriented data transfer;
- Class D: connectionless data transfer.

Cell Rate Decoupling

The cell rate decoupling is required for the physical layers that have synchronous cell timeslots (e.g., SONET, DS-3), whereas physical layers that have asynchronous cell timeslots do not require this function.

The transmitting end inserts unassigned and idle cells to the assigned cellstream (cells with valid payload) in the absence of the assigned cell and creates a continuous stream of assigned and idle cells. The receiving end extracts and discards the unassigned and idle cells from the cellstream leaving a noncontinuous stream of assigned cells. The rate at which the unassigned and idle cells are inserted or extracted depends on the bit rate of the assigned cell or the physical layer transmission rate. Specific header patterns are used for the unassigned and idle cells.

Cells Discrimination Based on Predefined Header Field Values

The predefined header field values are shown in Table 9.2. Metasignaling protocol uses the metasignaling cells for establishing and releasing VCC. For PVC, metasignaling is not used.

Specially designated OAM cells are used to exchange the VPC operation information between nodes (F4 flow). F4 flow OAM cells share the same VPI value as the user cell. The VCI = 3 indicates VP-level management functions between ATM nodes on a single VP link segment, and VCI = 4 indicates VP-level end-to-end management functions.

TABLE 9.2. The Predefined Header Field Values (UNI).
(*After:* GR-1113-CORE, Bellcore.)

Use	Value			
	Octet 1	Octet 2	Octet 3	Octet 4
Unassigned cell indication	aaaa0000	00000000	00000000	0000xxx0
Metasignaling	aaaayyyy	yyyy0000	00000000	00010a0c
General broadcast signaling	aaaayyyy	yyyy0000	00000000	00100aac
Point-to-point signaling	aaaayyyy	yyyy0000	00000000	01010aac
PHY cells	xxxx0000	00000000	00000000	0000xxx1
Segment OAM F4 flow cell	aaaaaaaa	aaaa0000	00000000	00110a0a
End-to-end OAM F4 flow cell	aaaaaaaa	aaaa0000	00000000	01000a0a
RM cells	aaaaaaaa	aaaa0000	00000000	01100a0a
Reserved	aaaaaaaa	aaaa0000	00000000	01110a0a
ILMI cell	aaaa0000	00000000	00000001	0000aaa0
SMDS cell	aaaa0000	00000000	00000000	11110a0a

a: use for the ATM layer function

x: do not care

y: any VPI value other than 00000000

c: The originating signaling entity sets the CLP bit to 0;
the network may change this value to 1.

Cell Discrimination Based on Payload Type Identifier Field Values

The PT field identifies user cells from nonuser cells. The PT values 0 to 3 define user cells. Values 2 and 3 indicate that congestion has been experienced in the network. The PT value of 4 identifies F5 flow OAM cells within a single link, while the PT value of 5 identifies end-to-end F5 flow OAM cells. See Table 9.3 for payload type indicator encoding.

Loss Priority Indication and Selective Cell Discarding

The CLP field is used for loss priority indication by the ATM end point. The ATM network equipment can selectively discard cells depending on the CLP value.

The ATM end equipment can set the CLP bit equal to zero or one when it first transmits an ATM cell. A higher priority cell will contain CLP = 0, and a lower priority cell will contain CLP = 1. When these cells enter the network, and the network experiences traffic congestion, all cells containing CLP = 1 may be discarded.

Traffic Shaping

Traffic shaping is a mechanism that achieves desired characteristics for the stream of cells entering into a VCC or a VPC. It alters the traffic characteristics of a stream of cells. Examples of traffic shaping are:

TABLE 9.3. Payload Type Indicator Encoding.
(*After:* GR-1113-CORE, Bellcore.)

PT coding	Meaning
000	User data cell, congestion not experienced, SDU-type = 0
001	User data cell, congestion not experienced, SDU-type = 1
010	User data cell, congestion experienced, SDU-type = 0
011	User data cell, congestion experienced, SDU-type = 1
100	Segment OAM F5 flow-related cell
101	End-to-end OAM F5 flow-related cell
110	Resource management
111	Reserved for future function

- Peak cell rate reduction;
- Burst length limiting;
- Suitably spacing cells in time.

ATM Layer Management (M-Plane)

Management functions at the UNI are reduced to a minimal set because there is some cooperation required between customer premises equipment and network equipment. The ATM layer management functions at the UNI deals mainly with the following management functions:

- Fault management;
- Performance management;
- Activation and deactivation;
- System management.

OAM type and function type are shown in Table 9.4.

OAM cell flows are used to exchange management information among nodes (including customer premises equipment). The F4 flows support segment or end-to-end (VP termination) management information at the VP level. VCI values 3 and 4 are used to identify F4 flows. The F5 flows support segment or end-to-end (VC ter-

TABLE 9.4. OAM Type and Function Type. (*After:* GR-1113-CORE, Bellcore.)

OAM type	Value	Function type	Value
Fault management	0001	Alarm indication signal (AIS)	0000
		Remote defect indication (RDI)	0001
		Continuity check	0100
Performance management	0010	Loopback	1000
		Forward monitoring	0000
Activation and deactivation	1000	Backward monitoring	0001
		Monitoring and reporting	0010
System management	1111	Performance monitoring	0000

mination) management information at the VC level. PT values 4 and 5 are used to identify F5 flows. At the ATM layer, the F4-F5 flows are carried over OAM cells. See Figure 9.5 for OAM cell flows at UNI. See Figure 9.6 for OAM cell format.

Fault Management

The fault management type contains the following management functions:

- Alarm surveillance;
- Connectivity verification;
- Invalid VPI/VCI detection.

The flow diagram of the ATM layer maintenance interaction is shown in Figure 9.7.

Alarm Surveillance

VPC/VCC failure indications at the public UNI are covered by the alarm surveillance function. The detection, generation, and propagation of such failure condition are carried by way of OAM cells.

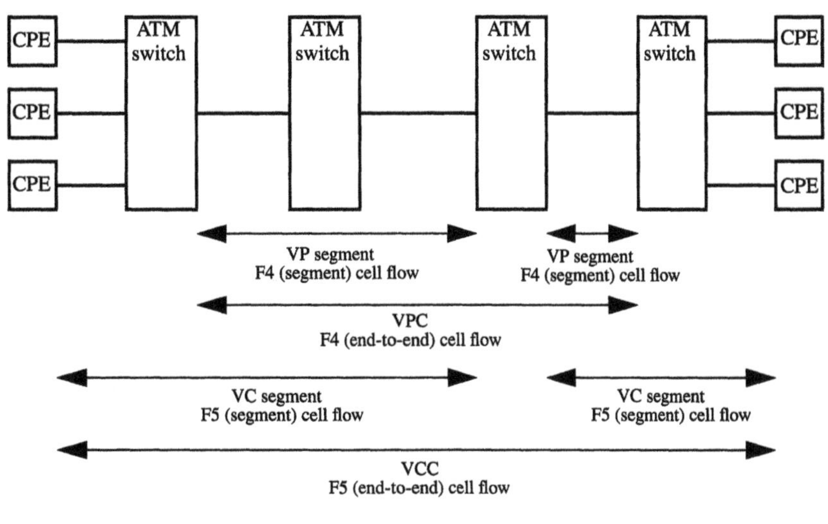

FIGURE 9.5. OAM cell flows at UNI.

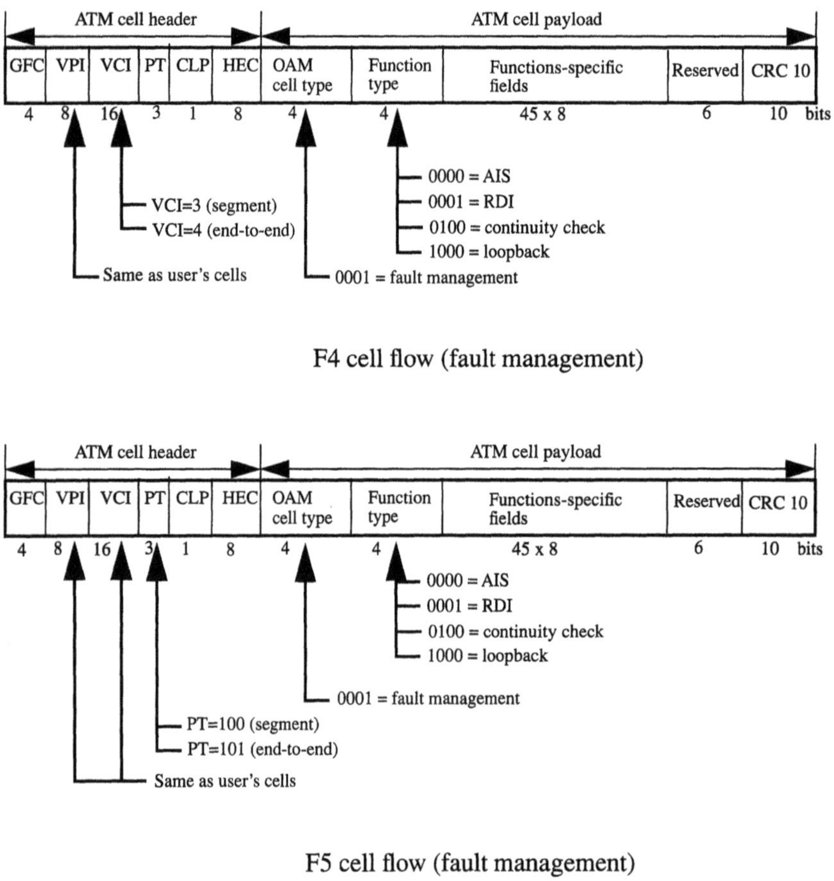

FIGURE 9.6. OAM cell format.

In the case of the SONET physical layer, AIS, RDI, and REI failure indication signals are used.

Upon detecting a VPC/VCC failure or by the notification of a physical layer failure, the VP/VC AIS (VP-AIS or VC-AIS) is generated by a VPC or VCC node at a connection point. VP-AIS or VC-AIS is generated to notify downstream VPC or VCC nodes that a failure has been detected upstream.

Upon receiving a VP-AIS/VC-AIS, the VPC/VCC end node returns a VP-RDI/VC-RDI. VP-RDI/VC-RDI is generated to alert the upstream nodes that a failure has been detected downstream. See Figure 9.8 for AIS/RDI OAM cells.

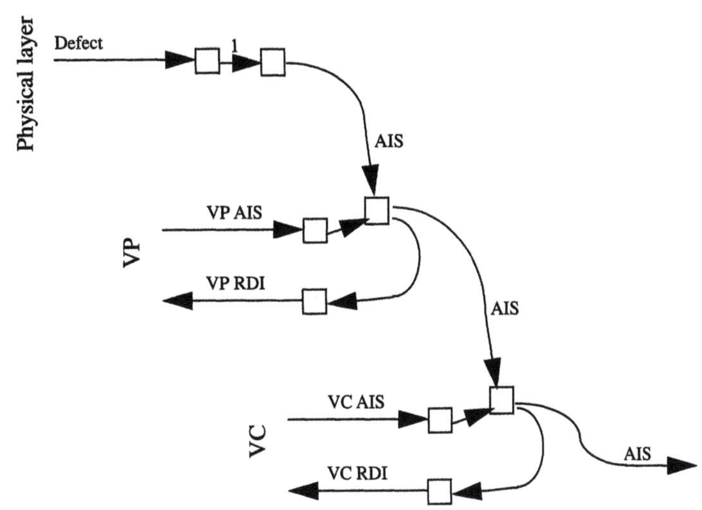

FIGURE 9.7. ATM layer maintenance interaction.

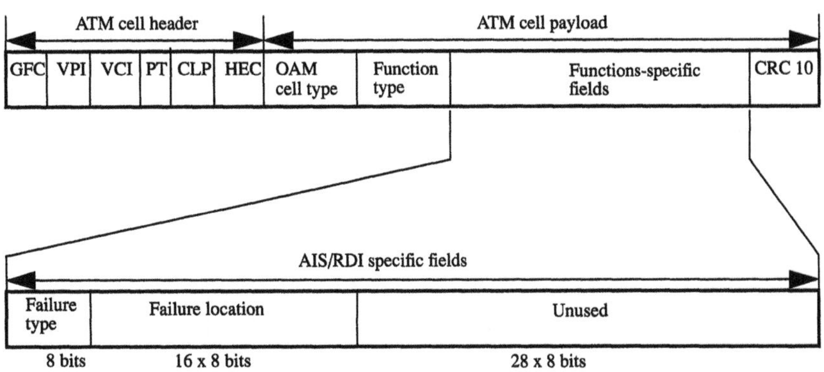

FIGURE 9.8. AIS/RDI OAM cell.

Connectivity Verification

Both VP and VC connections can be verified by OAM loopback cells. These cells are also used for failure localization and testing. While this loopback function is being performed, the VC or VP connections can remain in service. The OAM loopback cell format is shown in Figure 9.9.

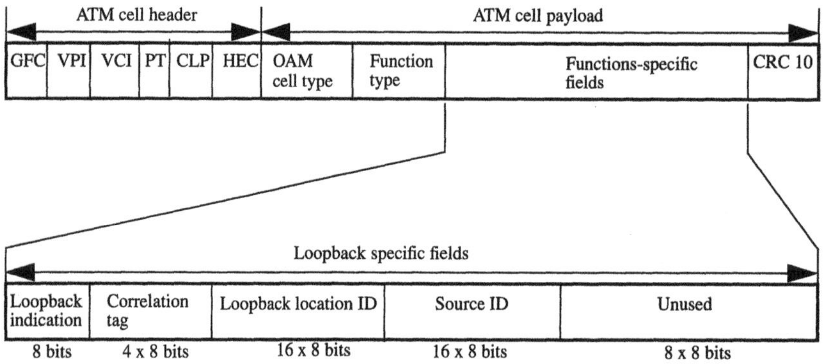

FIGURE 9.9. OAM loopback cell.

The four fields of the OAM loopback cells as defined by the standards committee are on the following pages.

Loopback Indication

This field is 8 bits wide. The receiver checks this field to determine whether the incoming cell is to be looped back. The seven most significant bits are always coded as 0. The least significant bit is 1 before the cell is looped back and 0 after the cell has been looped back.

Correlation Tag

This field is 32 bits wide. This field is used to correlate transmitting OAM cells with receiving OAM cells. The correlation is needed because multiple OAM fault management cells may be inserted in the same virtual connection at any given time. This field can be coded with any value, and the endpoint looping back the cell should not modify it.

Loopback Location ID

This field is 96 bits wide. This field identifies the loopback point along a virtual connection. The transmitter sets all ones to indicate the loopback location, which is the endpoint of the connection or segment.

Source ID

This field is 96 bits wide. This field is used to identify the originator of the loopback cell so the originator can identify the looped-back cell when it returns. This field is encoded with any value.

Invalid VPI/VCI Detection

At the receiving end, VPI/VCI values of ATM cells are checked for validity. If ATM cells contain invalid VPI/VCI, then these are discarded and layer management is informed.

Performance Management

The ATM layer management performs the performance monitoring and reporting and the activation and deactivation process in support of performance management. The performance management type contains the following management functions:

- Forward monitoring;
- Backward reporting;
- Monitoring and reporting.

See Figure 9.10 for the monitoring and reporting OAM cell.

Forward Monitoring, Monitoring and Reporting

The performance management monitoring procedure monitors items, such as bit error, loss/misinsertion of cells, and cell transfer delay. The functions-specific field of the forward monitoring and monitoring and reporting OAM cells contains the following fields:

- Monitoring cell sequence number (MSN);
- Total user cell (TUC) number;
- BIP-16;
- Time stamp (TS).

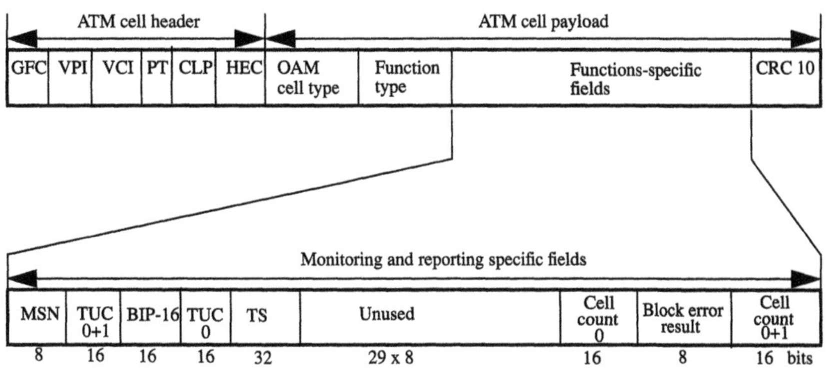

FIGURE 9.10. Monitoring and reporting OAM cell.

The MSN field is 8 bits wide. It indicates the sequence number of the forward monitoring, and monitoring and reporting messages. Each TUC field is 16 bits wide. It indicates the number of the last user cell in the block before the monitoring cell is inserted. The BIP-16 field contains the even parity error detection code computed over the information field of the block of user cells transmitted after the last monitoring cell. The TS field is 32 bits wide. It is used to represent the time at which the OAM cell was inserted.

Backward Reporting, Monitoring and Reporting

Performance management reporting procedure reports the monitoring results (i.e., error counts and loss/misinserted cell counts). The functions-specific field of the backward reporting and monitoring and reporting OAM cells contains the following fields:

- MSM;
- TUC number;
- BIP-16;
- TS;
- Block error result;
- Lost and misinserted cell count.

The MSN field is 8 bits wide. It indicates the sequence number of the forward monitoring, and monitoring and reporting messages. Each TUC field is 16 bits wide. It indicates the number of the last user cell in the block before the monitoring cell is inserted. The BIP-16 field contains the even parity error detection code computed over the information field of the block of user cells transmitted after the last monitoring cell. The TS field is 32 bits wide. It is used to represent the time at which the OAM cell was inserted.

The block error result field is 8 bits wide. It contains the number of errored parity bits in the BIP-16 code of the incoming monitoring cell. The lost and misinserted cell count field is 16 bits wide. It contains the count of lost or misinserted cells computed over the incoming monitoring block.

Activation and Deactivation

The ATM layer management performs the activation and deactivation process in support of performance management and continuity check. The activation and deactivation management type contains the following management functions:

- Performance monitoring;
- Continuity check.

See Figure 9.11 for activation and deactivation of an OAM cell.

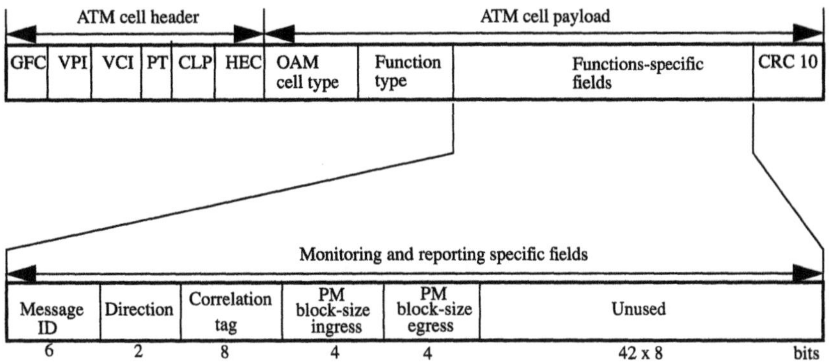

FIGURE 9.11. Activation and deactivation of an OAM cell.

System Management

The ATM layer management also deals with system management issues.

ATM Layer Operation

The ATM layer operation varies depending on its locality. It is somewhat different in an ATM end-station than in an ATM switch.

ATM End Station

The 48-octet information payload, service data unit (SDU), is passed from the AAL layer, together with various parameters, such as the values of the CLP and the PT fields, to the ATM layer. Then the ATM layer generates cells by pre-pending ATM header octets with the appropriate VPI/VCI values, and then passes the cell to the physical layer for transmission. The ATM layer also receives incoming ATM cells, on valid VPI/VCI values, and passes the SDU (i.e., the information payload) to the AAL layer, along with an indication of whether congestion has been experienced or not as specified by the PTI field. It exchanges an ATM cellstream with the physical layer. If it does not have any information to transmit, it inserts idle cells into the ATM cellstream.

An ATM end station can support multiple virtual connections at any given time. The order in which multiple virtual circuits need be serviced depends on the implementation. It can prioritize one connection to the other or serve in a round-robin fashion. Normally, the higher priority queues are served before the lower priority queues. ATM cells are transmitted in sequence as they are received from the AAL layer.

The ATM layer also polices the QoS for each connection as negotiated upon connection setup. There are several parameters specified during the connection request, such as cell loss rate, acceptable cell delay, and peak and average data rates. The network uses the requested QoS parameters for connection control. The request for connection can be rejected if there are insufficient resources (transmission bandwidth, buffers, etc.) existing across the network, from source to destination. The ATM end station may then repeat the request with lower QoS characteristics.

Each connection has its own committed QoS parameters. Traffic shaping at the source node and traffic policing at intermediate and end nodes are used to ensure that nodes meet, but do not exceed, their committed QoS parameters.

ATM Switch

When an ATM cell arrives at one of the input ports of an ATM switch, the ATM layer examines the VPI/VCI values in the ATM cell header and determines the output port to which the ATM cell should be transferred. The ATM cell is then forwarded to the output port with a new VPI/VCI value. The cell is then passed down to the physical layer of the outgoing port for transmission.

The PTI field is set by the ATM layer if congestion is experienced. The ATM layer also implements the traffic shaping and policing algorithms. Ordering is maintained for the cells from the same virtual circuit. The maximal end-to-end latencies are also met. The ATM layer has to store cells during cell processing, which requires buffers in the system. Multiple input ports may transmit cells to the same output port. Therefore, adequate buffering and other congestion control mechanisms are supported. The ATM layer also replicates cells to support multicast functions.

ATM Layer Design Example

The 53-octet ATM cells are carried over the ATM network. ATM end stations send and receive ATM cells on virtual connections. The virtual connections are established by using physical links and switching systems. A unique connection identifier is assigned to each virtual connection. The VPI/VCI of the ATM cell header is mapped into the connection identifier. This connection identifier is used by the receiving equipment to process the ATM cell. The connection identifier actually is a pointer to a memory block that contains all the necessary information to process incoming ATM cells. All cells belonging to a specific virtual connection follow the identical processing. The equipment that processes ATM cells is called the ATM processor. The ATM cell processing functions can be subdivided into two categories:

- Ingress cell processing;
- Egress cell processing.

See Figure 9.12 for a typical ATM line card block diagram. The CPU configures, controls, and monitors the status of the ATM processor. The context memory stores the connection parameters (information on how to process ATM virtual connections) for the ATM processor. The CPU-RAM stores program and temporary data. The DMA transfers block of data between the CPU-RAM and the ATM processor. The ATM processor interfaces between a PHY interface and a switch interface. The PHY interface board design example was discussed in the previous chapter. The ATM processor performs ATM cell processing and ATM cell routing functions.

Ingress Cell Processing

The ingress refers to ATM cells being transferred from the physical interface to the ATM switch. Figure 9.13 shows the ingress block diagram. The ingress cell processing in an ATM line card consists of the following steps:

- Incoming cell assembly from the physical layer;
- VPI/VCI (address) translation/compression;
- Context table lookup;
- User parameter control (UPC) processing (tagging, passing, discarding);
- Network parameter control (NPC) processing (tagging, passing, discarding);
- Cell counting (statistic, billing);
- OAM processing (fault, performance management);

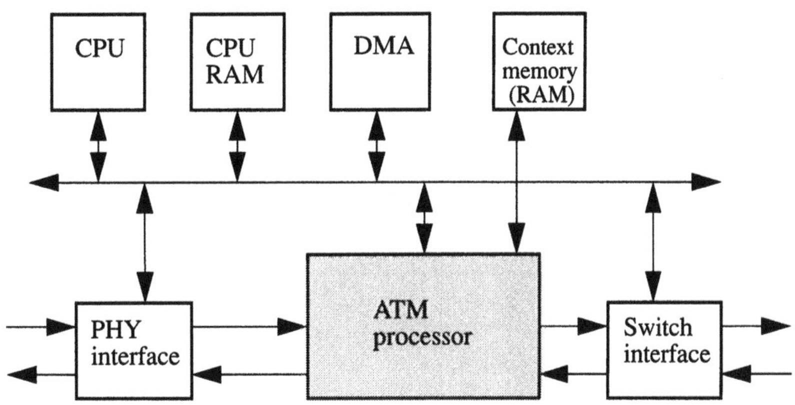

FIGURE 9.12. ATM line card block.

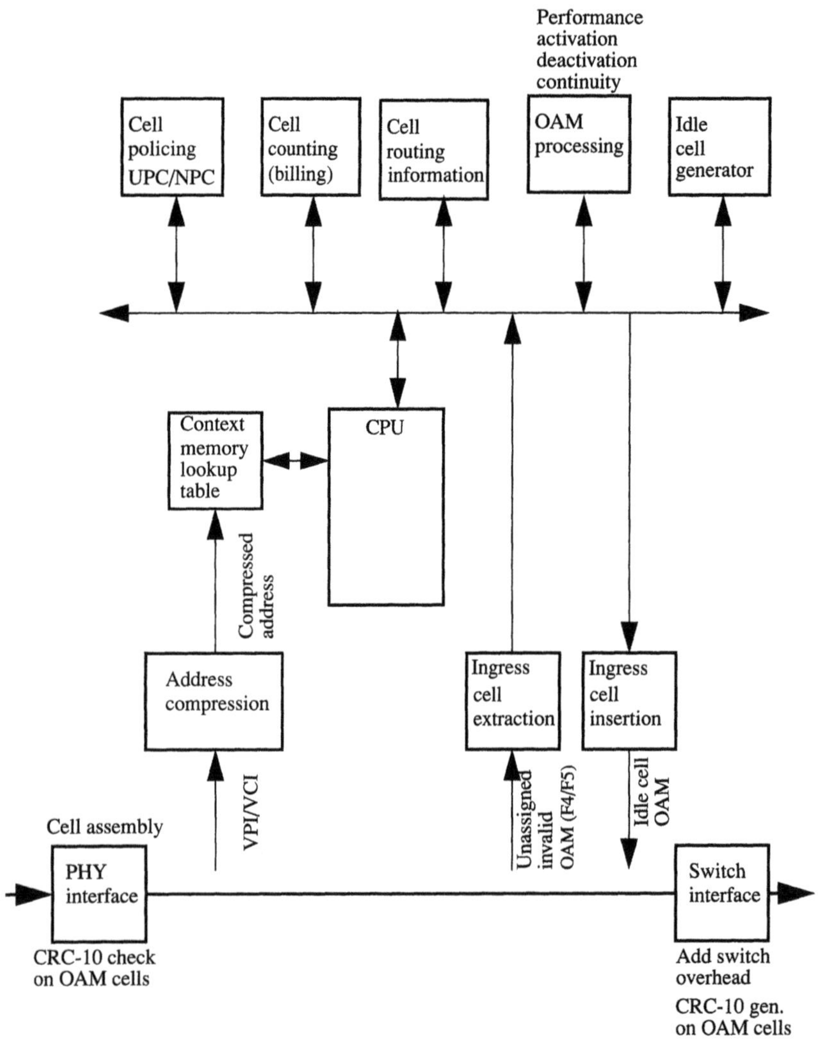

FIGURE 9.13. Ingress block.

- Cell extraction (OAM cells, idle cells, errored cells);
- Cell insertion (OAM cells, idle cells);
- Attaching switch overhead information for proper routing through ATM switch;
- Transfer of ATM cells to the ATM switch.

Ingress Cell Assembly Block (Physical Layer Interface)

The cell assembly block of the ingress section of the ATM processor interfaces with the physical layer device. This block extracts ATM cells from the FIFO located in the physical layer. The physical layer transfers cell data one octet at a time using the UTOPIA standard interface. The ingress block assembles octets into cells because the ATM processor works on a cell basis. These cells are held in a FIFO within the ingress section and are read by the ATM processor. Each cell is processed within one cell slot. The following equation represents a cell slot for OC-3 PHY interface:

$$\frac{(53\ \text{octets} \times 8\ \text{bits per octet})}{\text{OC3 line rate}} = \frac{53 \times 8}{155.52 \times 10^6}\text{sec} = 2.73\mu\text{s}$$

Similarly, a cell slot for DS-3 PHY interface is

$$\frac{(53\ \text{octets} \times 8\ \text{bits per octet})}{\text{DS3 line rate}} = \frac{53 \times 8}{44.736 \times 10^6}\text{sec} = 9.48\mu\text{s}$$

The idea is to process cells faster than the physical layer can provide cells for there to be enough gaps in the data flow in which to insert cells into the datastream. The header field is checked. Unassigned and invalid cells are discarded from the cell flow to provide cell rate decoupling. Both unassigned and invalid cells create gaps in the cell flow. Ingress cell assembly block is shown in Figure 9.14.

The PTI field is checked to identify different types of cells. Cells are counted according to CLP bits. OAM cells are counted separately.

If a cell is an OAM cell, then the CRC-10 field is checked, errors reported, and invalid cells discarded. If no error is found, then cell processing will be performed on the cell.

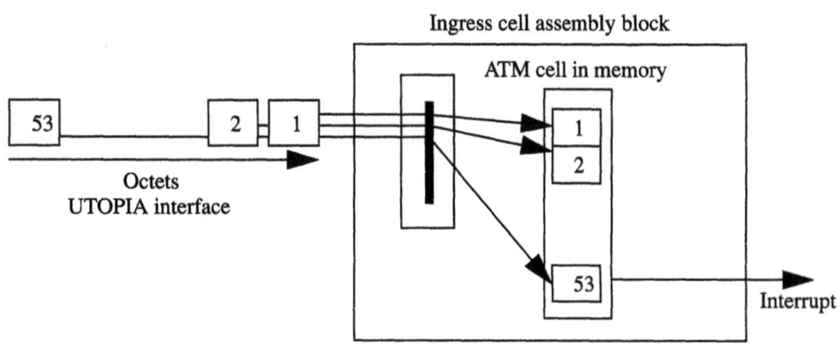

FIGURE 9.14. Ingress cell assembly block.

Address Compression Block

The purpose of the ingress address processing is to map the VPI/VCI fields of the incoming ATM cell header into a pointer to the entry in the context table that relates to the incoming cell's virtual connections (VPC/VCC). For VP switching, the VPI field is used to map into the pointer to the context table. If a cell has no valid connection then that cell is discarded. The ingress address compression block is shown in Figure 9.15.

The address compression is required because VPI/VCI fields represent up to 256 million different ATM connections. It is impossible to implement 256 million separate connection parameters in the memory. Therefore, address compression is required to support only a small set for the total combination; that is, active connections only.

The address compression mechanism is too complex for a few sentences to describe. A further discussion of this topic is found in Chapter 12 on VPI/VCI translation and compression.

Context Memory Lookup Table

The context memory lookup table contains all the necessary connection parameters for the incoming cells. The size of this table is determined by the number of active connections supported by the ATM processor at any given time. The context memory lookup table is shown in Figure 9.16.

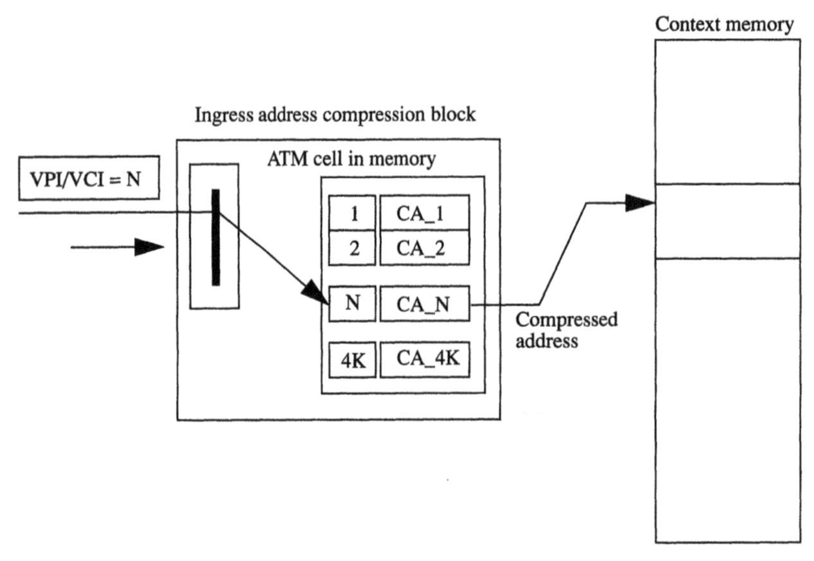

FIGURE 9.15. Ingress address compression block.

Some of the following information is stored in this table on a per connection basis:

- Ingress counters;
- Egress counters;
- Cell connection information;
- Switch overhead information;
- Cell policing information.

Cell Policing (UPC-User Parameter Control/NPC-Network Parameter Control) Block

The UPC/NPC functions are performed on a connection basis. In Chapter 14, UPC/NPC algorithms are discussed in more detail. The UPC/NPC block diagram is shown in Figure 9.17.

This algorithm detects cells that do not conform to the negotiated traffic parameters. Upon detecting a nonconforming cell, one of the following actions may be taken based on network congestion status:

- Cells may be tagged (CLP bit is set to 1) and passed to the switch;
- Cell may be discarded (removed from the cell flow);
- No action taken (statistics only).

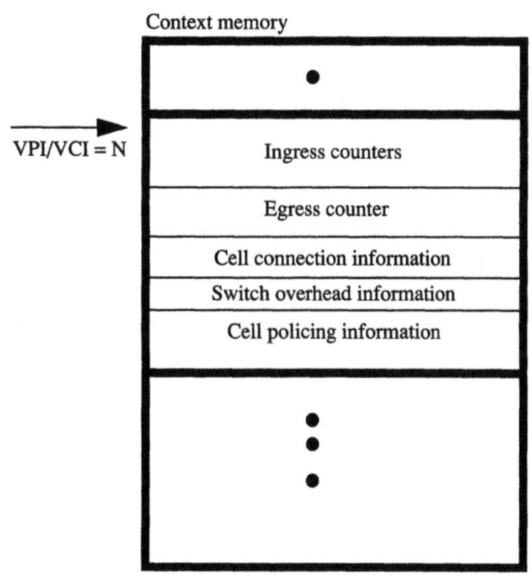

FIGURE 9.16. Context memory lookup table.

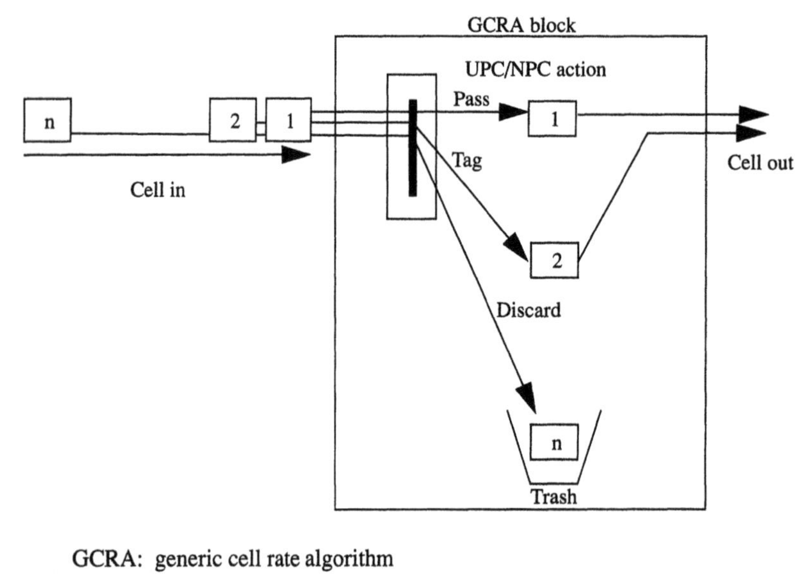

FIGURE 9.17. UPC/NPC block.

Cell Counting Block

Cells are counted for performance, billing, and statistical purposes. The counters are located in the context table on the basis of each connection. The counters are updated each time a cell arrives for that connection. When a cell arrives at the ATM cell processor, it is counted in stages of processing. Cells may be counted in the following stages:

- The cell assembly block counts the total number of cells that arrive in the ATM processor. This count may be further divided into user cell (CLP=0), user cell (CLP=1), OAM cell (CLP=0), or OAM cell (CLP=1).
- The cell policing block updates the counters when a cell is tagged. A separate counter is updated when a cell is discarded.
- The cells with errors are counted separately.

The ingress cell counting block is shown in Figure 9.18.

Cell Routing Information Block

The cell routing information block adds the proper routing information into the ATM cell header. The HEC field is also added to the ATM cell header. The switch overhead appending is shown in Figure 9.19.

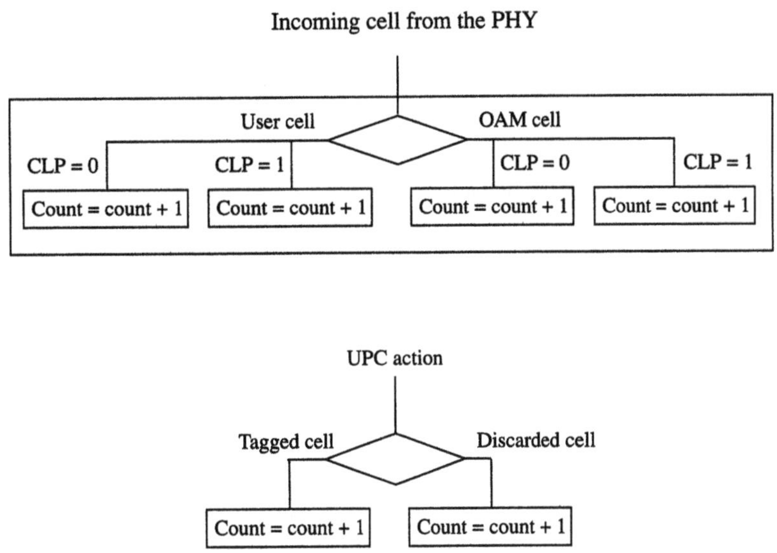

Note: A separate count is maintained for each VPI/VCI number; that is, each connection.

FIGURE 9.18. Ingress cell counting block.

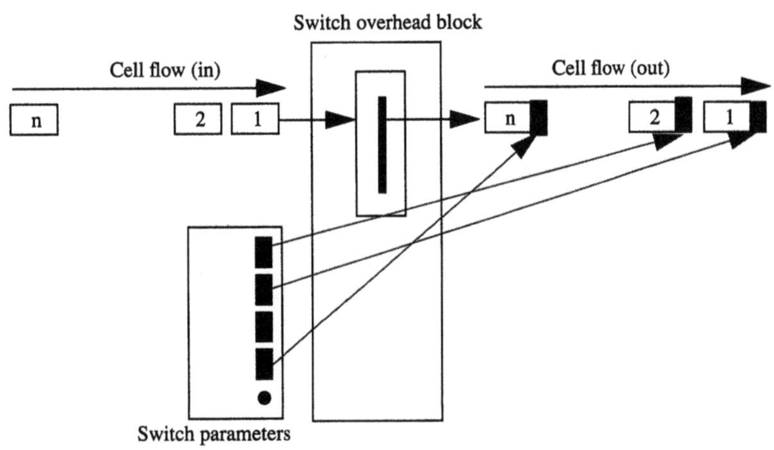

FIGURE 9.19. Cell routing information block.

OAM Processing Block

The function of the OAM processing block is to process the OAM cells received. After analyzing an OAM cell, this block may perform the following tasks:

- Performance checking;
- Activation and deactivation;
- Continuity checking (loopbacks);
- Fault information gathering.

Idle Cell Generation Block

When there is a gap in the cellstream, the idle cell generated by this block is inserted into the cellstream.

Ingress Cell Extraction Block

This block extracts cells from the following type of cells from the cellstream and stores them in the extraction buffer. They are queued by the processor to analyze:

- OAM cells;
- Unassigned cells;
- Invalid cells;
- Errored cells;
- Nonconforming cells.

The ingress cell extraction block is shown in Figure 9.20.

Ingress Cell Insertion Block

The ATM processor uses gaps in the cell flow to insert cells into the stream. The cell insertion is further controlled by a general cell rate algorithm (GCRA) to ensure that the cell insertion rate stays within capacity. Cells are inserted at a rate that does not violate the traffic characteristics of the ATM switch. The types of cells that can be inserted are:

- OAM cells (AIS cells, continuity check cells, performance monitoring cells);
- Idle cells;
- ILMI cells.

The ingress cell insertion block is shown in Figure 9.21.

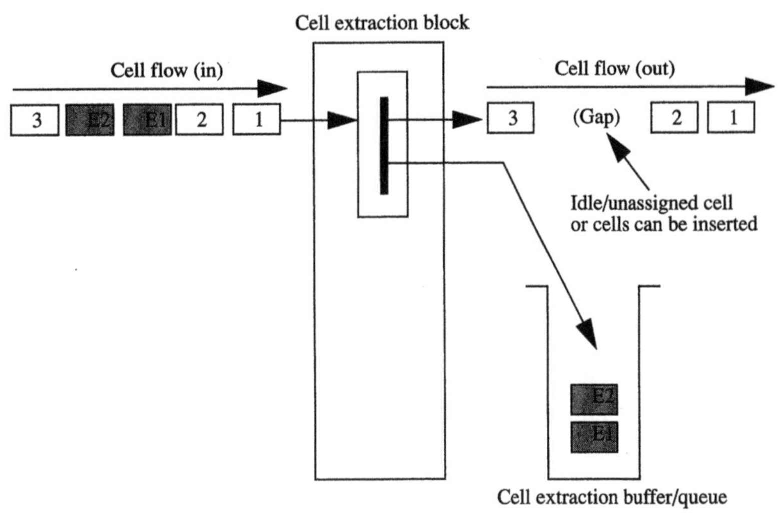

FIGURE 9.20. Cell extraction block.

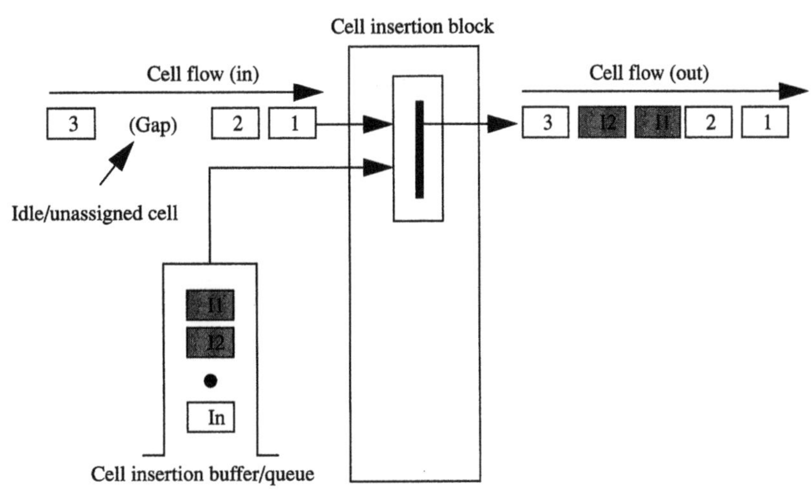

FIGURE 9.21. Cell insertion block.

Switch Interface Block

This block interfaces with the ATM switch. ATM cells with the appropriate switch overhead are serialized and passed to the switch for proper routing. The switch interface block is shown in Figure 9.22.

Egress Cell Processing

The egress refers to ATM cells being transferred from the ATM switch to the physical interface. Figure 9.23 shows the egress block diagram.

The egress cell processing in an ATM line card consists of the following steps:

- Incoming cell assembly from the ATM switch;
- Multicast identifier translation and compression;
- Context table lookup;
- OAM processing;
- Cell extraction;
- Cell insertion;
- Address translation (replace old VPI/VCI);
- Transfer ATM cells to the physical layer.

Egress Cell Assembly Block

The cell assembly block of the egress section of the ATM processor interfaces with the ATM switch. This block extracts ATM cells from the FIFO located in the ATM switch. The ATM switch transfers cell data one octet at a time using the UTOPIA standard interface. The egress block assembles octets into cells since the ATM pro-

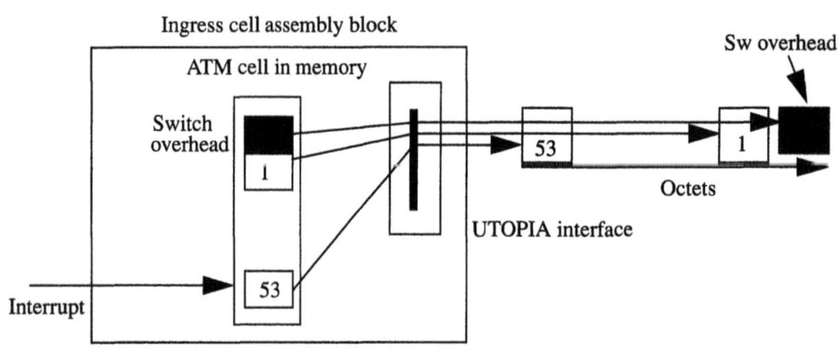

FIGURE 9.22. Ingress switch interface block.

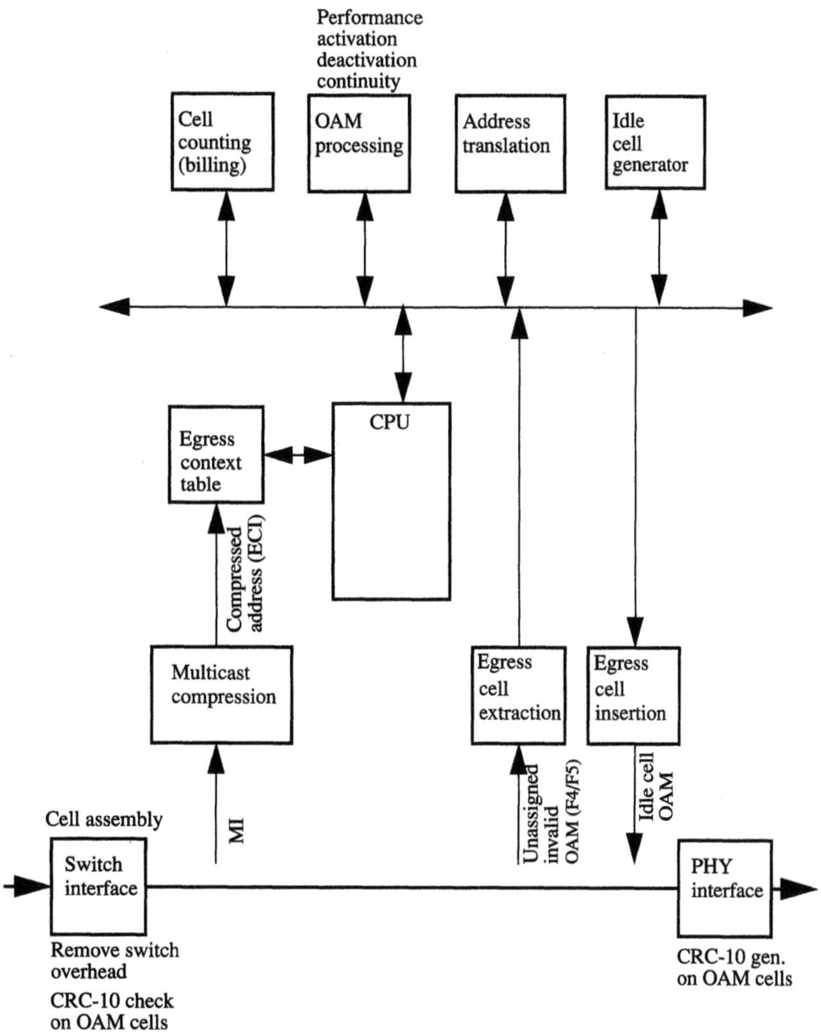

FIGURE 9.23. Egress block.

cessor works on a cell basis. These cells are held in a FIFO within the egress section and are read by the ATM processor. Each cell is processed within one cell slot.

The switch overhead field is checked. Unassigned and invalid cells are discarded from the cell flow to provide cell rate decoupling. Counters are counted according to CLP bits. OAM cells are counted separately. If a cell is an OAM cell, then the CRC-10 field is checked, errors reported, and invalid cells discarded. If no error is found, then cell processing is performed on the cell. The egress cell assembly block is shown in Figure 9.24.

Multicast Translation Block

The switch overhead contains the egress connection identifier (ECI) or multicast connection identifier (MCI). Multicasting means to copy cells to multiple physical links. The table that translates the multicast indentifer is called the multicast translation table. This table provides multiple ECIs for a single MCI.

These ECIs are used to look up the context table for connection parameters. The multicast address translation block is shown in Figure 9.25.

Egress Context Table

The egress context table is pointed directly by the ECI and contains all the connection parameters for each connection. The size of this table is determined by the number of active connections supported by the ATM processor at any given time. The following information is stored in this table on a per connection basis:

- Egress counters;
- OAM parameters;
- VPC/VCC connection types.

Egress Cell Counting Block

Cells are counted for performance and statistical purposes. The counters are located in the context table on the basis of each connection. The counters are updated each

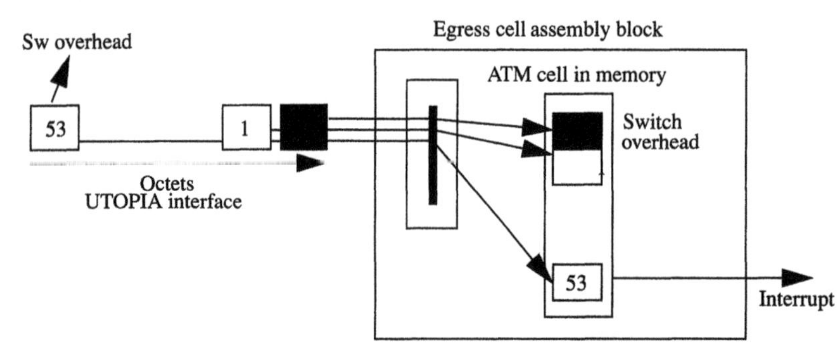

FIGURE 9.24. Egress cell assembly block.

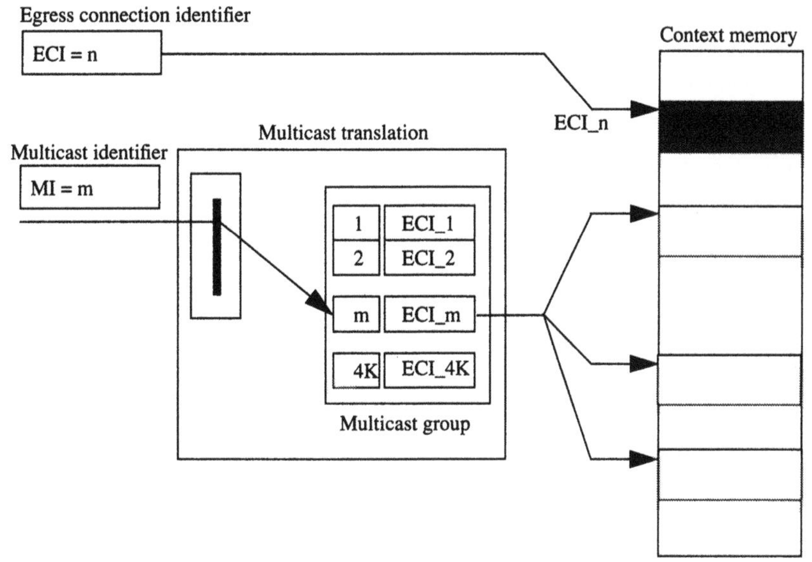

FIGURE 9.25. Multicast address translation block.

time a cell arrives for that connection. When a cell arrives at the ATM cell processor, it is counted:

- The cell assembly block counts the total number of cells that arrive in the ATM processor. This count may be further divided into user cell (CLP=0), user cell (CLP=1), OAM cell (CLP=0), and OAM cell (CLP=1).
- The cells with errors are counted separately.

The egress cell counting block is shown in Figure 9.26.

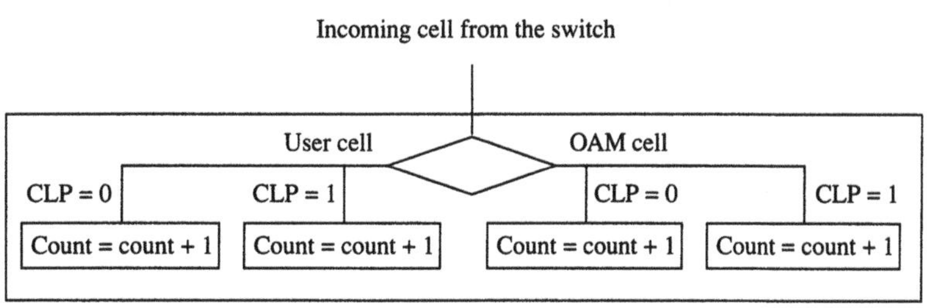

FIGURE 9.26. Egress cell counting block.

OAM Processing Block

The function of the OAM processing block is to process the OAM cells received. After analyzing an OAM cell, this block may perform the following tasks:

- Performance checking;
- Activation and deactivation;
- Continuity checking (loopbacks);
- Fault information gathering.

Idle Cell Generation Block

When there is gap in the cellstream, the idle cell generated by this block is inserted into the cellstream.

Egress Cell Extraction Block

This block extracts cells from the following types of cells from the cellstream and stores them in the extraction buffer, which is queued by the processor to analyze:

- OAM cells;
- Unassigned cells;
- Invalid cells;
- Errored cells.

Egress Cell Insertion Block

Cells are inserted in a gap that is created by the ATM processor. The cell insertion is further controlled by a GCRA to ensure that the cell insertion rate stays within capacity. The cells that are to be inserted are stored in a FIFO and the insertion rate is regulated. The types of cells that can be inserted are:

- OAM cells (RDI cells, continuity check cells, performance monitoring cells);
- Idle cells;
- ILMI cells.

Egress Address Translation Block

The ATM cell header is modified in this block (new VPI/VCI is added). The egress address translation block is shown in Figure 9.27.

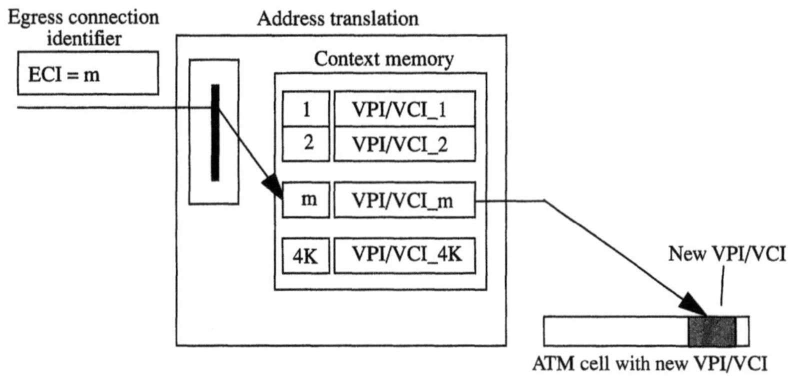

FIGURE 9.27. Egress address translation block.

Egress Physical Layer Interface Block

Cells are transferred to the physical layer device one octet at a time. The interface is the UTOPIA standard interface. The multi-PHY interface is also handled in this block. If a cell is an OAM cell, then the CRC-10 field is added. The egress physical layer interface block is shown in Figure 9.28.

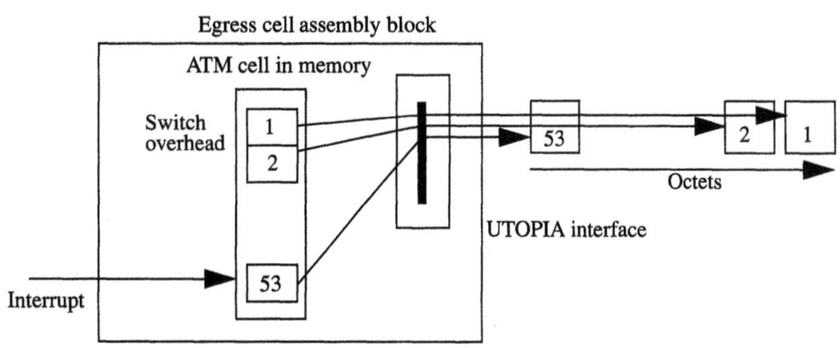

FIGURE 9.28. Egress physical layer interface block.

CHAPTER 10

ATM Adaptation Layer

Background Information

Many different types of services with many different characteristics are offered by the data network. Some of these services are as follows:

- Interactive audio (e.g., telephone);
- Audio distribution (e.g., radio);
- Audio retrieval (e.g., audio library);
- Interactive video (e.g., video conference);
- Video distribution (e.g., television, distance learning);
- Video retrieval (e.g., video on demand);
- Frame relay interworking;
- Response time critical transaction (e.g., airline reservations, banking, stock market);
- Interactive text, data, and image transfer (e.g., credit card verification, banking transaction);
- Text, data, and image messaging (e.g., e-mail, fax);
- Text, data, and image distribution (e.g., news, weather report);
- Text, data, and image retrieval (e.g., file transfer);
- LAN (e.g., LAN emulation);
- Remote terminal (e.g., home office).

Telephone (voice/speech) is a full-duplex service and requires low bandwidth. It is a constant bit rate service and delay sensitive. Videoconference is a full-duplex service and requires high bandwidth with constant or variable bit rate service and is also delay sensitive. Cable TV requires low delay variance. It is a constant bit rate service that requires high bandwidth. Data, image, and fax are the same type of service. All of these vary in characteristics, but they are delay insensitive. Some of these services require high bandwidth.

ATM Adaptation Layer Types

ATM has been designed to support all types of traffic on the same transmission and switching fabrics. The ATM adaptation layer (AAL) protocols generate the traffic that is carried in the ATM cells. It is accomplished by performing a convergence function, which is to package the information (data) received from the higher layers into the 48-byte information payload that is then passed down to the ATM layer for transmission. In the receiving end, the AAL receives the information payloads passed up from the ATM layer and puts these into the form expected by the higher layer. Therefore, the AAL process is the most important feature of the ATM communications process. The adaptation process clearly varies depending on the type of traffic that is carried—in fact, the AALs are broken into five different AAL protocols, each of which is designed to optimally carry one of the four classes of traffic. See Figure 10.1 for the ATM adaptation layer diagram.

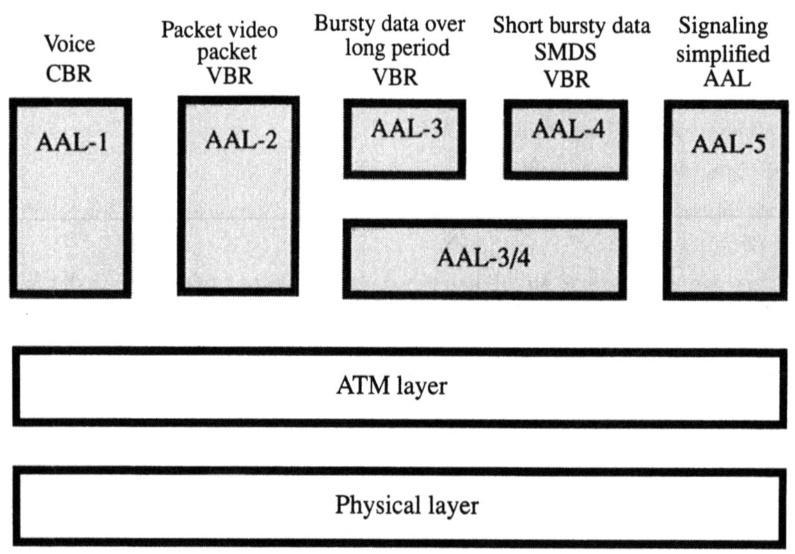

FIGURE 10.1. ATM adaptation layer.

The AALs are designed to carry the following four basic types of traffic:

1. Constant-bit rate (CBR), connection-oriented, synchronous traffic (e.g., uncompressed voice or video);
2. Variable-bit rate (VBR), connection-oriented, synchronous traffic (e.g., compressed voice and video);
3. VBR, connection-oriented, asynchronous traffic (X.25, frame relay, etc.);
4. VBR, connectionless packet data (LAN traffic, SMDS, etc.).

See Figure 10.2 for the types of ATM services descriptions.

It is normally assumed that only the classes of traffic with which each of the AALs is associated will support that particular AAL (e.g., AAL 1 will normally offer class A service and will carry class A traffic). Nevertheless, once a connection is established across the ATM network, the network does not care what kind of traffic is carried in the ATM cells or how that information might be packaged in them.

Several different types of AAL can be supported by a particular ATM end node. These types of ATM end nodes ensure some local means of associating a particular instance of an AAL entity with its corresponding ATM layer entity.

All types of AAL protocols share a common function called *segmentation and reassembly (SAR)*. The SAR is the process by which a higher layer *service data unit (SDU)* that is passed to an AAL is segmented into units that can be carried in the 48-octet information payloads. For instance, if a large data packet is passed to an AAL, the AAL will have to divide it into multiple 48-octet segments before it

	AAL type 1	AAL type 2	AAL type 3/4,5	AAL type 3/4
Asynchronous/ synchronous	Synchronous		Asynchronous	
Bit rate	Constant	Variable		
Connection mode	Connection-oriented			Connection-less
Examples	Uncompressed voice or video	Compressed voice or video	X.25, frame relay	LAN traffic, SMDS

FIGURE 10.2. Types of ATM services.

passes them to the ATM layer. The receiving ATM end node reassembles these small pieces back into the original SDU before passing it to the higher layer.

Sufficient information is passed with the SDU segments to allow the destination ATM end node to reassemble the SDU segments back into a packet. Because the ATM cells are fixed in size, all AAL types do perform some form of SAR. The exact means by which SAR is performed, however, vary by AAL type.

Refer to Bellcore documents GR-1113-CORE and GR-1115-CORE and the ATM Forum document AF-BICI-0013.003 for more information.

AAL Type 1

Class A AAL type 1 is connection-oriented. It performs the functions necessary to adapt CBR services to ATM layer services. This service requires a strong timing relation between the source and the destination. The following are examples of such traffic that could be transported across the public or private ATM UNI:

- PCM-encoded voice traffic;
- CBR video;
- Transport of DS-1 links;
- Transport of DS-3 links.

The AAL-1 connections are established between two AAL-1 user entities over pre-established point-to-point connections. The AAL-1 accepts data from a higher layer at a CBR (e.g., a fixed number of bits at a regular time interval), and delivers to the destination at the same rate. The AAL protocol layer places the received data in an AAL-1 SAR-protocol data unit (SAR-PDU). This 48 byte SAR-PDU is then passed down to the ATM layer as an information field for an ATM cell. CBR supports the following two types of data transfer:

1. Structured data transfer;
2. Unstructured data transfer.

Structured Data Transfer

Structured data transfer allows CBR services to transport octet-based information, such as *n* x 64 kbps ISDN channels.

Unstructured Data Transfer

Unstructured data transfer allows CBR services to transport a bitstream with an associated service clock, such as DS-1 and DS-3.

See Figure 10.3 for the CBR protocol stack, and see Figure 10.4 for the AAL-1 circuit emulation diagram.

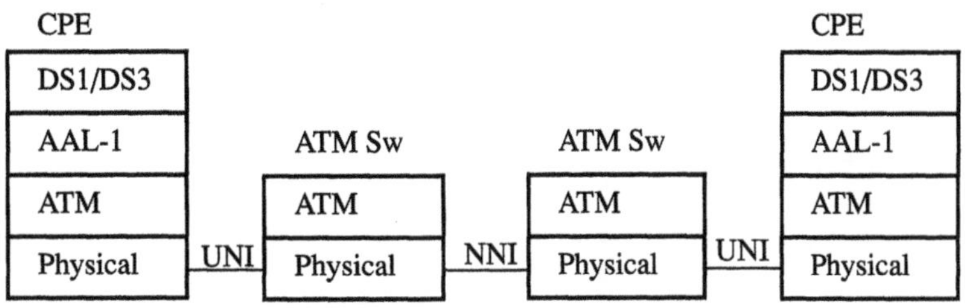

FIGURE 10.3. CBR protocol stack.

PDU Structure and Encoding

The AAL-1 SAR-PDU consists of a one-octet header followed by a 47-octet payload (AAL-1 SAR-SDU). This 48-octet AAL-1 SAR-PDU is the ATM cell payload (ATM-SDU). See Figure 10.5 for the AAL-1 SAR-PDU structure.

AAL-1 SAR-PDU Header Field

The AAL-1 SAR-PDU header is shown in Figure 10.6.

The header contains the sequence number (SN) and the sequence number protection (SNP) fields.

Sequence Number

The first four bits of the header octet is the SN. The SN field is subdivided into two fields which are:

Convergence Sublayer Indicator (CSI): A 1-bit field is used for the transport of the residual time stamp (RTS) value for asynchronous clock recovery.

Sequence Count (SC): A 3-bit field contains a binary encoded sequence count value between peer AAL-1 CS-entities. This 3-bit counter is cycled through for each SAR-PDU passed to the ATM layer and helps detect the deletion or misinsertion of cells.

Sequence Number Protection

The last four bits of the header octet is the SNP, and this field is subdivided into two fields:

CRC control: This is the 3-bit CRC, calculated over the SN field and appended to the 4-bit SN field to produce a 7-bit code word. The CRC calculation does not include the SAR-PDU payload, because a bit error in a single PCM sample of voice traffic is generally unnoticeable. The bit errors in CBR data are acceptable.

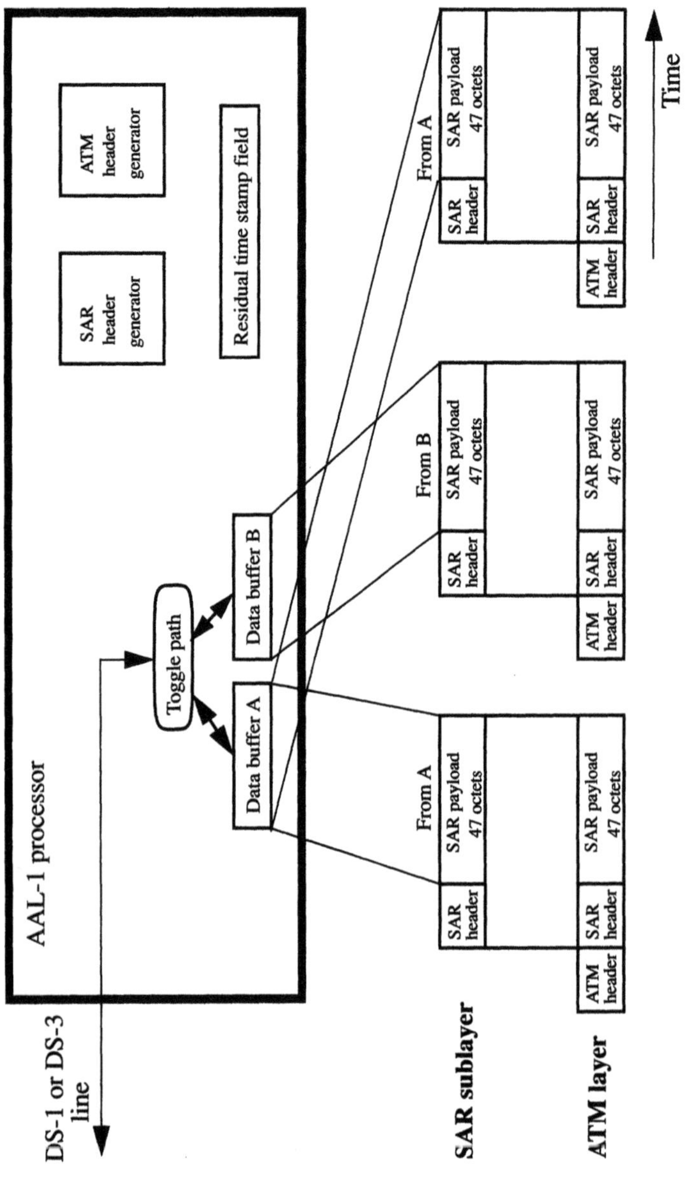

FIGURE 10.4. AAL-1 circuit emulation.

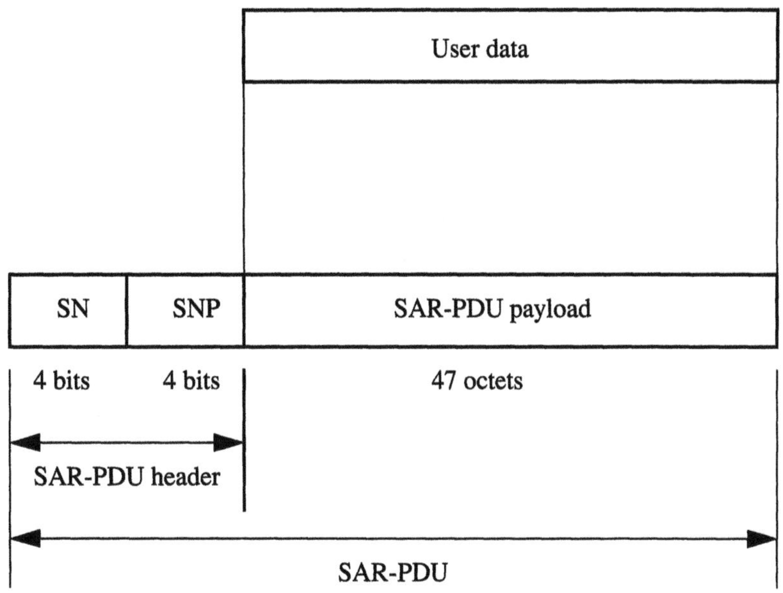

FIGURE 10.5. AAL-1 SAR-PDU structure.

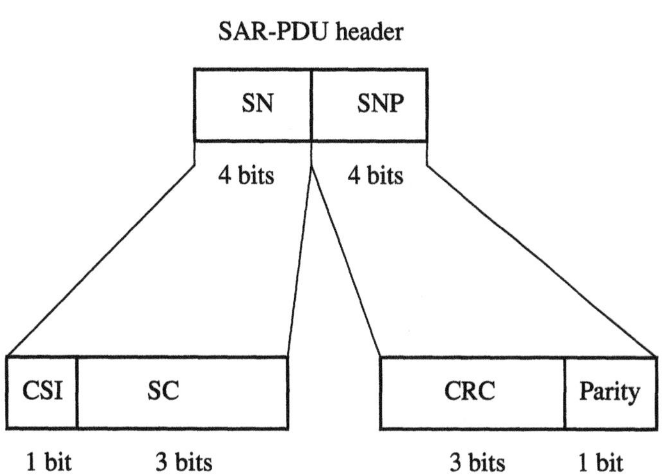

FIGURE 10.6. AAL-1 SAR-PDU header field.

Parity: This field is an even parity check over the 7-bit code word generated by the CRC processing. This bit occupies the last bit position of the SNP field.

SAR-PDU Payload (Information Field)

The information carried in an AAL-1 SAR-PDU payload field is for unrestricted data transfer service. Although the AAL-1 SAR-PDU can carry up to 47 octets of information, it is not a requirement that all 47 octets be used. The amount of information that may be carried is determined by how much packetization delay can be tolerated; that is, the number of samples that can be stored at the source and transported in one cell.

Synchronization

There is no procedure for maintaining synchronization from the source to the destination because AAL-1 does not make allowances for a clock to be carried. The ATM network was thought to be synchronous. The indication was that the ATM layer will perform the function of seeing that the ATM cells are sent and received at the same rate across the ATM network. This should occur when the samples are passed to the AAL-1 layer. Possibly, a clock could be generated at the destination from the cell interarrival time by itself. This might be accomplished through a phased locked loop and an elastic buffer.

Circuit Emulation Service

Circuit emulation (CES) allows transport of CBR traffic (DS-1, DS-3, etc.) using ATM technology. Please refer to ITU-T specification I.363, ATM Forum documents AF-SAAA-0032.000 and AF-PHY-0016.000, and Bellcore specification GR-1113-CORE for more information. The CBR service may have an asynchronous service clock and when it does, the transmitting AAL-1 entity generates timing information for an asynchronous service clock recovery process. The receiving AAL-1 entity is responsible for generating AAL-1 AIS when in the starvation condition. It is also responsible for generating AAL-1 with dummy bits (47 x 8) when the ATM cell carrying the AAL-1 SAR-PDU is lost.

The ITU-T has proposed the following two techniques for synchronous links (e.g., SONET):

1. Synchronous residual time stamp (SRTS);
2. Adaptive clock recovery.

Synchronous Residual Time Stamp

SRTS requires a networkwide common reference clock. This technique uses special synchronization information that is generated and sent to the destination by the source AAL, with which, together with a clock extracted from the physical layer, the destination clock is generated. SRTS gives better control over the wander introduced into the CES circuit.

SRTS measures the asynchronous DS-1/DS-3 clock frequency against a networkwide reference clock and transmits the difference value called RTS through the CSI bit of the AAL-1 SAR-PDU header. At the receiver, this difference is added to the networkwide reference clock to recover the asynchronous source clock frequency. The SRTS process for DS-1 CES is shown in Figure 10.7.

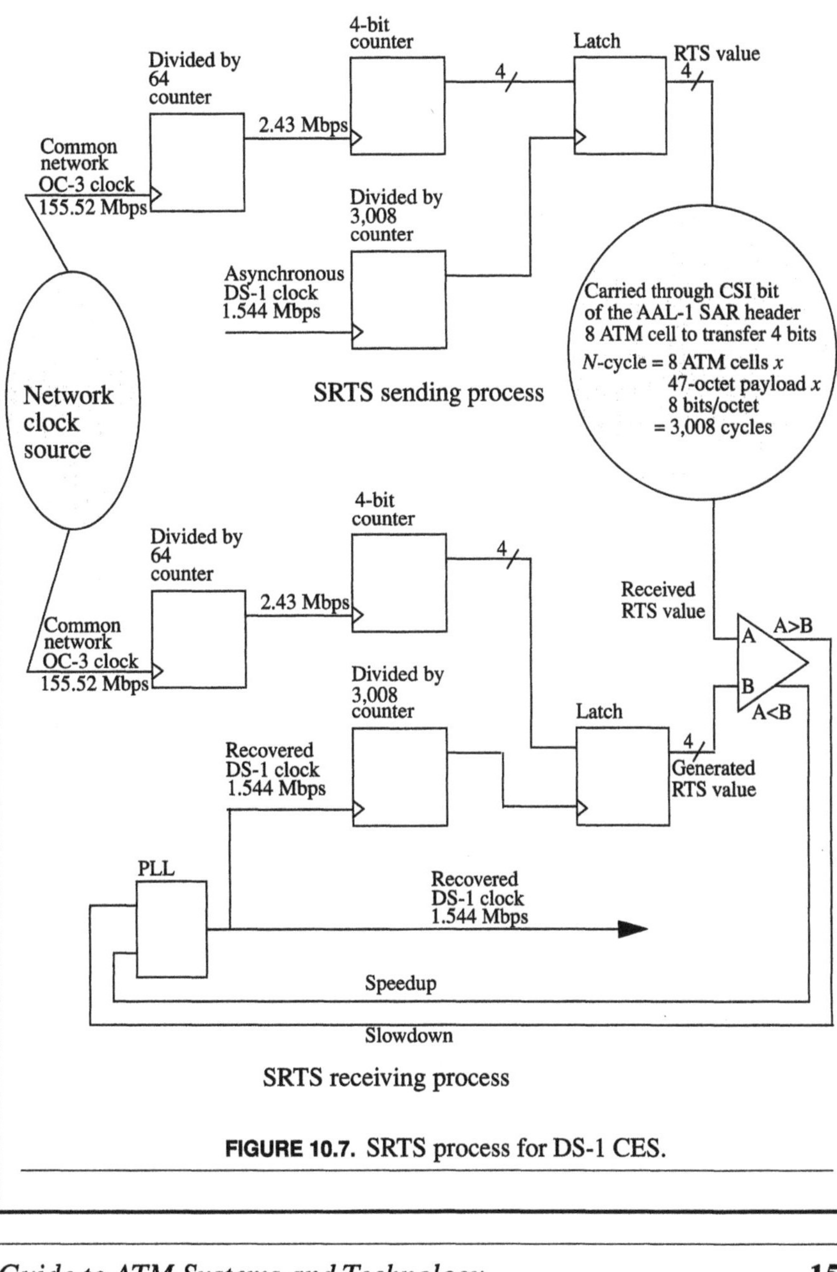

FIGURE 10.7. SRTS process for DS-1 CES.

Adaptive Clock Recovery

The adaptive clock recovery technique does not require a networkwide reference clock to recover the asynchronous source clock. The depth of the reassembly buffer at the receiver can be monitored to regain the asynchronous clock frequency. When the buffer depth increases, the received data from the buffer are read out faster to maintain a constant buffer depth. Similarly, when the buffer depth decreases, the received data from the buffer are read out more slowly to maintain the same buffer depth.

AAL Type 2

Class B traffic is connection-oriented traffic, and though the bit rate may be variable, it requires a close timing relationship between the source and the destination. Examples of such traffic are:

- VBR voice (compressed and packetized voice);
- VBR video (compressed and packetized video).

AAL-2 packetizes the data passed to it from the convergence sublayer into the SAR sublayer as an SAR payload. An SAR payload, together with a header and a trailer, form the SAR-PDU. This SAR-PDU then passes to the ATM layer for transmission.

The data being passed to the ATM layer from the higher layer arrive at a fixed interval and must be passed on the same way. In this case, the amount of data being passed at each transfer may vary. The amount of data may even exceed the capacity of a single cell. This would require the AAL to segment the data at the source into multiple cells and reassemble the cells at the destination before passing it to the higher layer. See Figure 10.8 for the AAL-2 CS-PDU segmentation diagram.

AAL-2 SAR-PDU Header and Trailer Field

The AAL-2 SAR-PDU header and trailer is shown in Figure 10.9.

Sequence Number

The 4-bit counter is cycled through for each SAR-PDU passed to the ATM layer. It helps to maintain the sequence of cells.

Information Type

The IT field can take on one of three values:

1. *Beginning of message (BOM)*: Indicates that the SAR-PDU contains the first set of data passed to it by the higher layer;

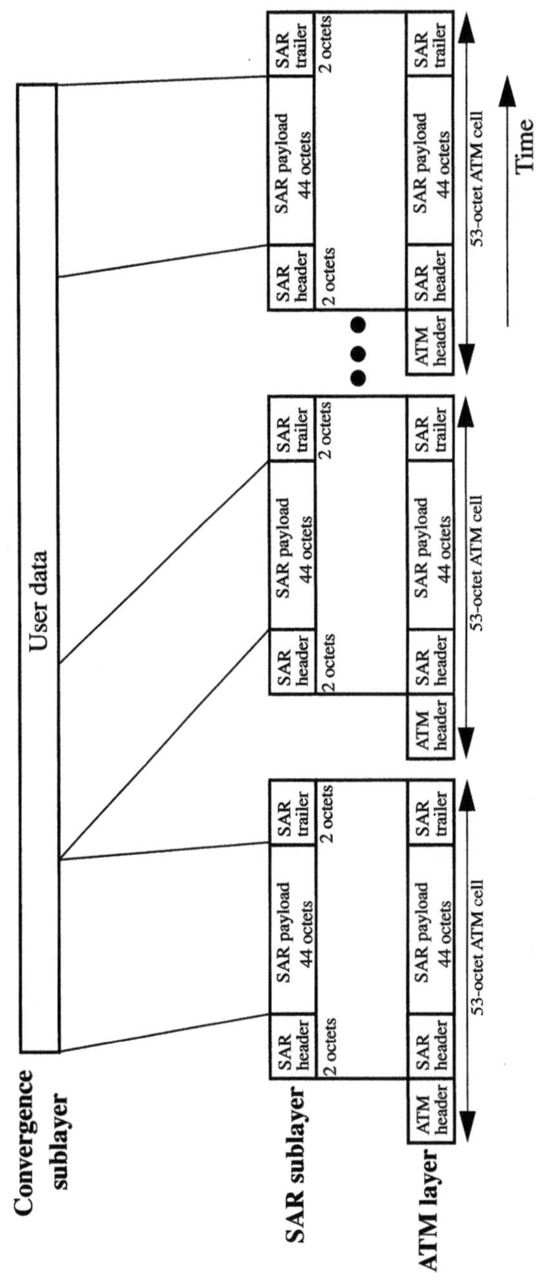

FIGURE 10.8. AAL Type-2 CS-PDU segmentation.

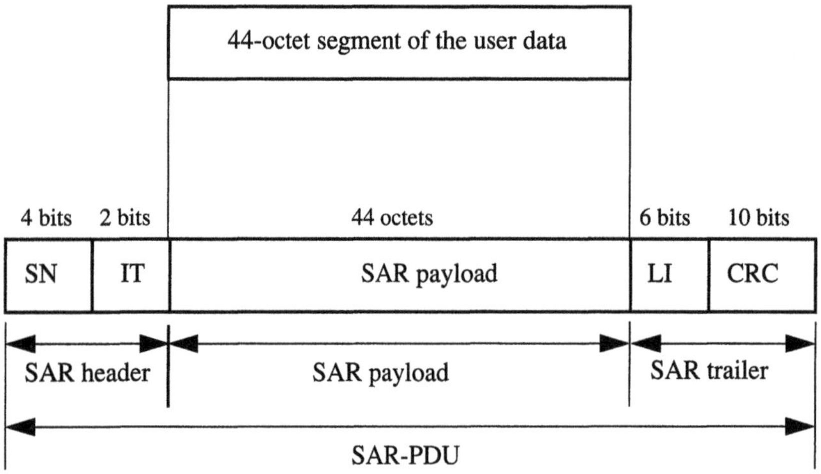

FIGURE 10.9. AAL-2 SAR-PDU.

2. *Continuation of message (COM)*: Indicates that the SAR-PDU contains additional data that are neither the beginning nor the end of the SAR-PDU;
3. *End of message (EOM)*: Indicates the last set of data to be sent.

If data passed to the AAL must be segmented into multiple cells at the source, likewise, they must be reassembled at the destination before being passed to the higher layer. The IT field identifies the different parts of the information. The AAL may request the ATM layer to set the CLP bit differently for BOM, COM, and EOM because class B traffic very often contains much more important data in the first few bytes of a particular transfer than in the last.

Information Payload

The user data are carried in this field. There is no restriction on the number of octets being used. The length indicator (LI) field is used to indicate how much of the payload field contains useful information.

Length Indicator

The LI field indicates how much of the payload contains actual data. For example, in an EOM segment, only a portion of the payload may contain useful data.

Cyclic Redundancy Code

Class B data have no tolerance for error; therefore, the CRC field is used to check for and correct bit errors over the entire SAR-PDU. It is important that class B data not be corrupted, since they are often generated by compression algorithms (e.g., differential coding of video signals), in which a single bit error will cause drastic error as a perceptible glitch.

AAL-2 Multiplexing Example

When we talk on the telephone, one normally listens and the other speaks. We also introduce lots of void (silence with background noise) into the system. We can take advantage of these situations in ATM technology. We do not need to transmit voice cells during the silence period. We can easily transmit cells from an active source instead of cells from a silence source. With proper voice compression and a silence suppression algorithm, one can achieve 3:1 bandwidth savings. Figure 10.10 shows an AAL-2 voice multiplexing block diagram.

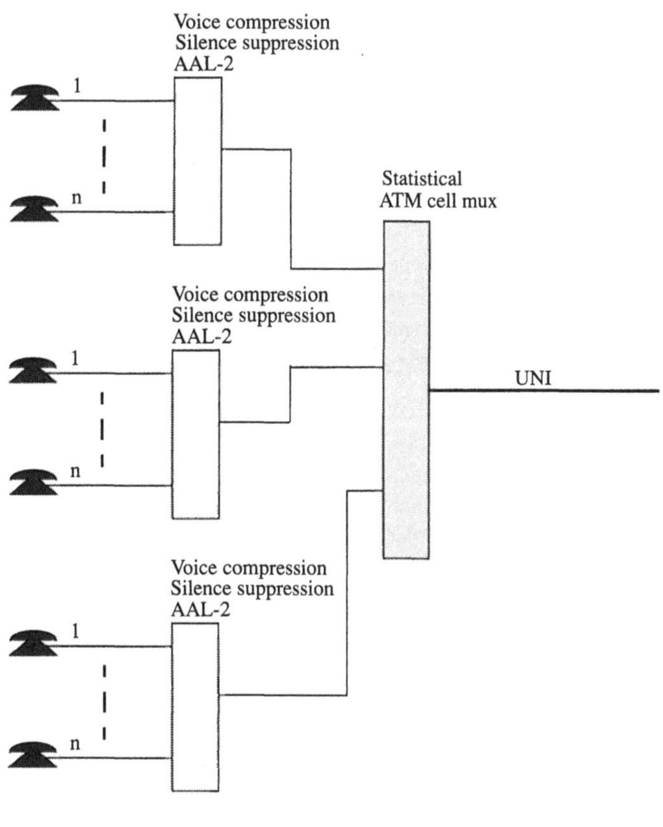

FIGURE 10.10. AAL-2 multiplexing.

AAL Type 3/4

The AAL-3/4 started out to be two different AAL specifications. These were merged and came to be known as AAL-3/4. AAL-3/4 is designed to transport connectionless variable-length frames (up to 65,536 octets) with error detection capability, over a pre-established connection. Connectionless traffic has no timing relationship between source and destination. Examples of such traffic are:

- X.25;
- Frame relay;
- TCP/IP.

However, there is no restriction on the type of traffic that can be carried by the AAL-3/4 PDUs once the connection is set up.

Connectionless, variable rate data, where there is no timing relationship between the source and destination, can be carried using AAL-3/4. An example of this type of traffic would be the connectionless packet data carried by LANs.

Connectionless LAN traffic can be carried by AAL-3/4 once a connection is established. The difference between the true connectionless and the AAL-3/4 connectionless modes of operation is that in the former, no connections are established before to ATM cell transfer.

In connectionless traffic, each ATM cell carries the destination address of the recipient so that the cell can be routed to the proper destination without any connection having been established earlier, such as SMDS traffic.

There are situations where connectionless service may be better suited than connection-oriented service. For example, a node may wish to eliminate the delay in setting up a connection with a destination, with which it communicates only occasionally, and save the overhead of setting up and clearing a circuit for only a short period.

ATM is basically a connection-oriented service. It seems unusual to discuss connectionless traffic in a connection-oriented environment. Connectionless ATM cells can be routed across reserved paths (VPI/VCI). The connectionless ATM cells are routed to the appropriate destinations by using the addressing information contained in the ATM cell header. VPI/VCI fields of each ATM cell are used as an index to a routing table for proper destination address.

The most popular way of carrying connectionless traffic is to establish a connection by using connection-oriented procedures, then connectionless ATM cells can be transferred through this established connection.

AAL-3/4 operates in the following modes:

- Message mode service;
- Streaming mode service (pipeline).

Fixed- or variable-size SDUs are passed to the AAL and transmitted in a single CS-PDU in the message mode, and a single SDU is passed to the AAL and transmitted in multiple CS-PDUs in the streaming mode, as when pieces of the SDU are received. The streaming mode of operation reduces the latency experienced by the packet.

See Figure 10.11 for the AAL type-3/4 CS-PDU segmentation.

AAL-3/4 Functions

The following functions are performed by the AAL-3/4:

- Preservation of the AAL-3/4 SDU;
- AAL-3/4 SDU segmentation;
- AAL-3/4 SDU reassembly;
- Error detection and handling;
- Multiplexing and demultiplexing;
- AAL-3/4 SDU abort;
- Pipelining.

Protocol Data Unit Structure

The protocol data structure for the AAL-3/4 contains two parts:

1. Convergence sublayer (CS);
2. SAR sublayer.

The convergence functions are performed by the CS sublayers that are required to map the higher layer information into the SAR sublayer and then the ATM layer.

CS-PDU Structure and Encoding

The CS-PDU is composed of variable-length information units passed to the CS sublayer from the higher layer. This CS-PDU packet is then passed to the SAR sublayer. The SAR sublayer segments the CS-PDU into its own 48-octet SAR-PDUs and passes them to the ATM layer for transmission.

The structure of the AAL-3/4 CS-PDU is shown in Figure 10.12; the fields are described below.

Common Part Indicator (CPI)

This field is 1-octet wide. This field is used to identify the message type that indicates the usage of the other fields in the CS-PDU. It represents the number of counting units (bits or octets) for the BASize and length field.

Beginning Tag (BTag)

This field is one octet wide. This field associates the headers and trailers of a CS-PDU and ensures that all SAR-PDUs have been received correctly. The AAL

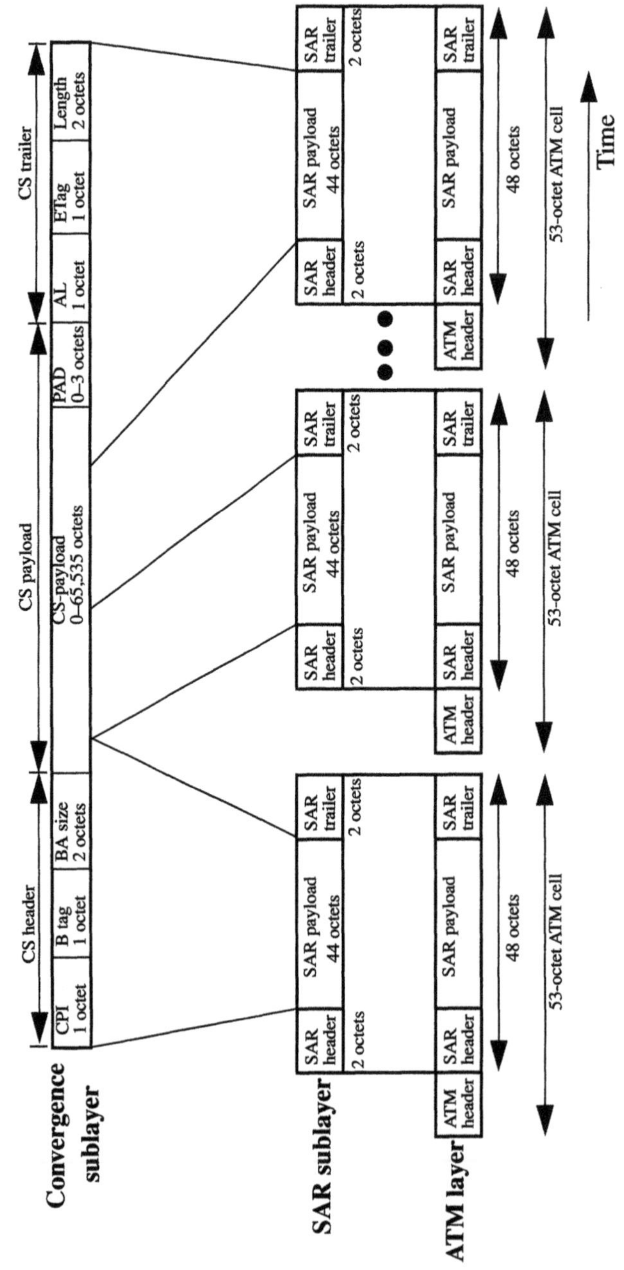

FIGURE 10.11. AAL-3/4 CS-PDU segmentation.

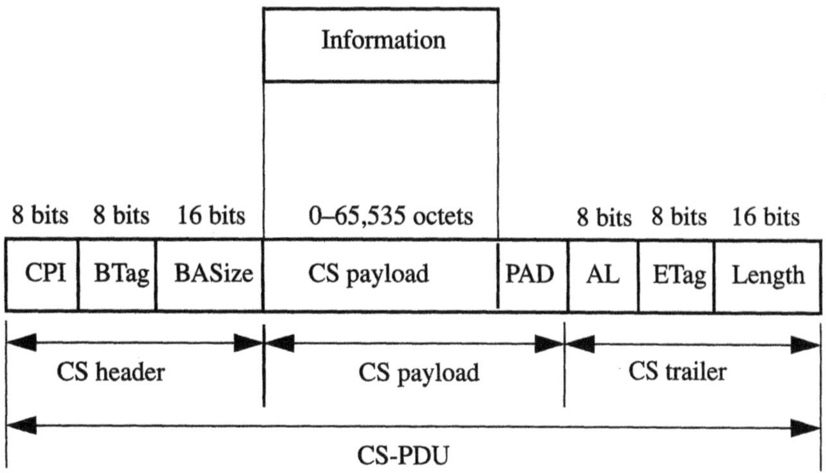

FIGURE 10.12. AAL-3/4 CS-PDU structure.

verifies whether the BTag and ETag of CS-PDUs match. If there is a mismatch, then it is declared that the CS-PDU is corrupted. This field increments as module 256 with successive CS-PDUs, and the same value is also placed in the end tag (ETag).

Buffer Allocation Size Indication (BASize)

This field two octets wide contains the value that indicates the maximal size of the buffer required to reassemble the CS-PDU. The receiving AAL monitors this field.

This is the length of the CS-PDU in the message mode. It contains some maximal value in the streaming mode because the length of the SDU being transmitted is unknown to the transmitting AAL.

Information Payload

This field contains the variable-length user information. It is octet aligned and the maximal size of the information payload (the minimal being zero) is the maximal value of the BASize field (65,536) times the size of the counting unit, which is indicated by the CPI field (typically, in octets).

Pad

The pad field is used to ensure that the total length of the CS-PDU payload field is 4-octet aligned. It is used as filler and does not convey additional information. The pad field is typically filled with zeros, and its length will vary between 0 and 3 octets.

Alignment (AL)

This 1-octet AL field is used to ensure that the CS-PDU trailer field is 4-octet aligned. It is used as filler and does not convey additional information. The AL field is filled with zeros.

End Tag (ETag)

This field is one octet wide and is used to detect error conditions through correlation with the BTag field in the header.

Length

This field is two octets wide and set to the actual length of the information field. The length value is less than or equal to the value of the BASize field. This field is also used to detect misassembled CS-PDU.

SAR-PDU Structure and Encoding

The CS-PDU is passed from the CS sublayer to the SAR sublayer, where it is segmented into one or more SAR-PDUs.

The structure of the AAL-3/4 SAR-PDU is shown in Figure 10.13.

The AAL-3/4 SAR-PDU fields are described below.

Segment Type

This two-bit field indicates whether a particular SAR-PDU is carrying the first section of a CS-PDU, BOM, an intermediate section, COM, the last section, EOM, or

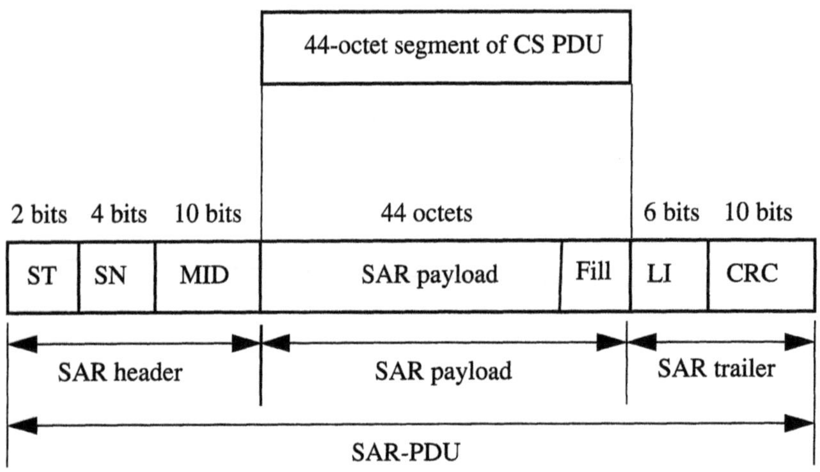

FIGURE 10.13. AAL-3/4 SAR-PDU structure.

whether the entire CS-PDU is in the single SAR-PDU, SSM. The encoding of the ST field is shown in Table 10.1.

Sequence Number (SN)

This field is four bits wide. It is used to identify the sequential position (modulo 16), and detect the loss or misordering of SAR-PDUs. The value of this field increments for each successive SAR-PDU for a CS-PDU.

Multiplexing Identification (MID)

This field is10-bit wide and assists in the interleaving of SAR-PDUs from different CS-PDUs and reassembly of these SAR-PDUs. In other words, it is used to differentiate between multiple CS-PDUs that are transmitting at the same time over the same ATM connection. A single CS-PDU uses the same MID value on all of the SAR-PDUs.

Information Payload

This field is 44 octets wide. The information payload of an AAL-3/4 SAR-PDU contains the segmented pieces of a CS-PDU.

Fill

This field contains the octets required to fill up the 44 octets of the Information Payload. This field is encoded with all zeros.

Length Indicator (LI)

This 6-bit length indicator field is used to indicate the length, in octets, the number of octets of the 44 octets of payload, which actually contain user data. An EOM or SSM may not contain 44 octets of data (e.g., a section of a CS-PDU). The value of the LI is in multiples of four octets.

Cyclic Redundancy Check Code

The 10-bit CRC covers the entire SAR-PDU and is used for bit errors detection.

TABLE 10.1. Encoding of the ST Field

ST	Encoding	Usage	LI
BOM	10	Beginning of message	44
COM	00	Continuation of message	44
EOM	01	End of message	4, 8, ..., 44
SSM	11	Single segment message	8, 12, ..., 44

AAL-3/4 Multiplexing Example

Figure 10.14 shows an ATM terminal that has two inputs with two variable-length packets arriving simultaneously destined for a single ATM VC using the AAL-3/4 protocol. BTag for Y=29, MID values are chosen between 4 and 8, and SN values start with 0.

AAL Type 5

The AAL-5, also known as the "simple and efficient AAL" (SEAL), was adopted by both the ATM Forum and by ANSI. Concern over the complexity of AAL-3/4 leads to the definition of a new AAL protocol, namely AAL-5. The AAL-5 was introduced to perform a subset of AAL-3/4 functions. Consequently, it is simpler than AAL-3/4.

The effective payload of an ATM cell with AAL-3/4 is, at best, equal to 83% (44/53), whereas, it is 90.5% (48/50) for AAL-5.

AAL-5 is designed to transport connectionless variable-length frames (up to 65,536 octets) with error detection capability, over only a pre-established connection. Connectionless traffic has no timing relationship between source and destination. Examples of such traffic are:

- X.25;
- Frame relay;
- TCP/IP.

However, there is no restriction on the type of traffic that can be carried by the AAL-5 PDUs once the connection is set up.

Connectionless LAN traffic is carried by AAL-5 once a connection is established. The difference between the true connectionless and the AAL-5 connectionless modes of operation is that in the former, no connections are established before ATM cell transfer.

The most popular way of carrying connectionless traffic is to establish a connection by using connection-oriented procedures. Then connectionless ATM cells can be transferred through this established connection.

See Figure 10.15 for the AAL-5 CS-PDU segmentation.

AAL-5 Functions

The following functions are performed by the AAL-5:

- Preservation of the AAL-5 SDU;
- AAL-5 SDU segmentation;
- AAL-5 SDU reassembly;

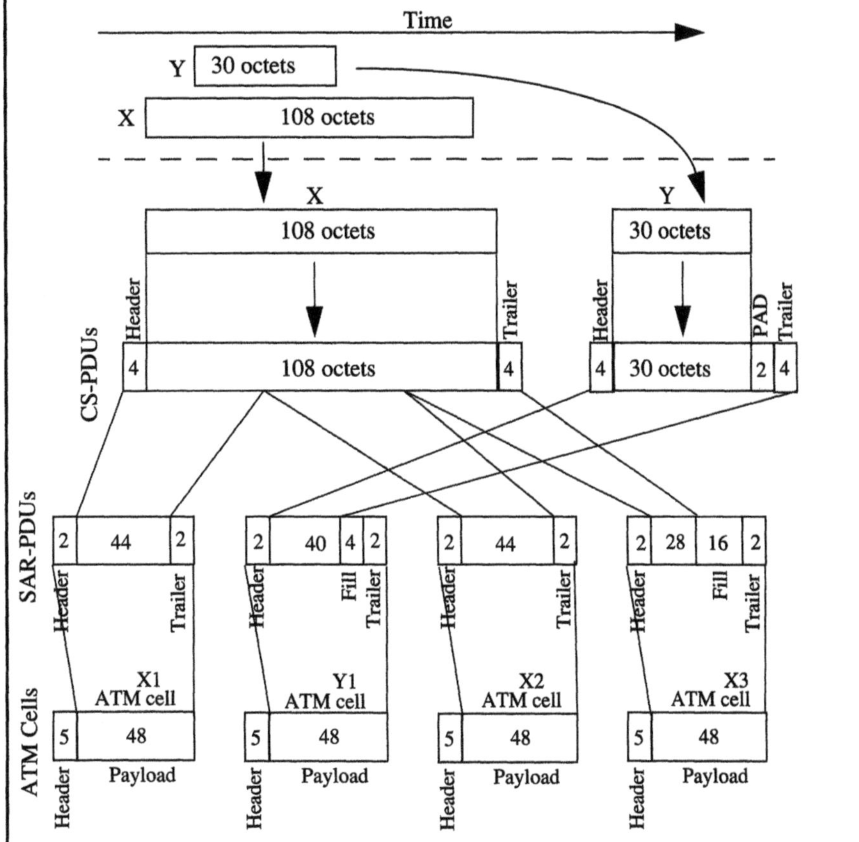

CS-PDUs

	X	Y
BTag	28	29
BA Size	116	40
PAD	none	2
ETag	28	29
Length	108	30

SAR-PDUs

	X1	X2	X3	Y1
ST	2	0	1	3
SN	0	1	2	0
MID	4	4	4	8
Fill	none	none	16	4
LI	44	44	28	40

ATM cell transmission order

Transmission order	X1	Y1	X2	X3

FIGURE 10.14. AAL-3/4 multiplexing example.

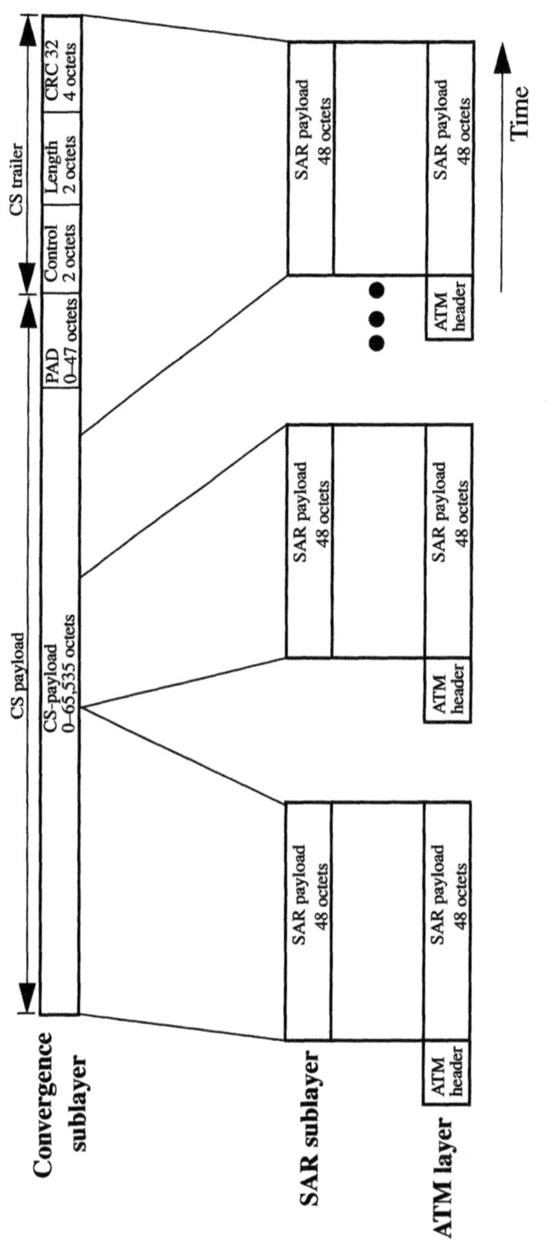

FIGURE 10.15. AAL-5 CS-PDU segmentation.

- Error detection and handling;
- AAL-5 SDU recovery;
- Reset;
- Connection parameter update.

Protocol Data Unit Structure

The PDU structure for the AAL-5 contains two parts:

1. CS;
2. SAR sublayer.

The convergence functions are performed by the CS sublayers that are required to map the higher layer information into the SAR sublayer and then the ATM layer. The AAL-5 SAR-PDU is simply 48 octets of data with no overhead of SAR-PDU headers or trailers. See Figure 10.16 for the AAL-5 SAR-PDU header field description.

CS-PDU Structure and Encoding

AAL-5 CS-PDU is passed to and from an SAR function. The fields of the AAL-5 CS-PDU are described below.

Information Payload

This field contains AAL-5 SDU. The AAL-5 information payload is octet aligned and can vary from 0 to 65,535 octets.

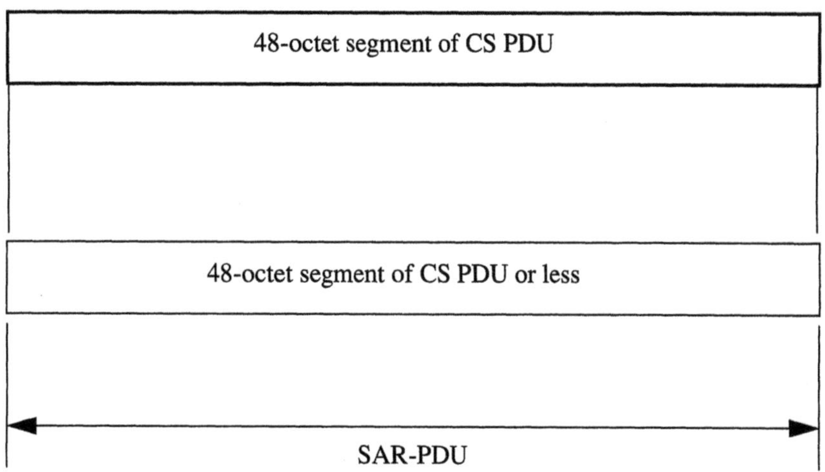

FIGURE 10.16. AAL-5 SAR-PDU header field.

Pad

This field varies between 0 and 47 octets. It is octet aligned to ensure that the CS-PDU is a multiple of 48 octets.

Control

This field is two octets wide and is reserved to support future AAL functions. it is coded with all zeros.

Length

This two-octet length indicator field is used to indicate the length, in octets, of the information payload.

CRC-32

This four-octet field contains the result of the CRC-32 calculation over the CS-PDU and is used to detect bit errors.

Because of AAL-5's simplicity, it is relatively easy to implement. The CS-PDU is divided into 48-byte segments, which are then passed to the ATM layer. The SAR-PDUs do not require the length indicator because the CS-PDU is 48-octet-aligned. The SAR sublayer does not detect the beginning or end of a CS-PDU reception. The PTI field of the ATM cell header determines the end of the AAL-5 SAR-PDU. The AAL-5 sets the SDU-type to 1 for the last cell of a CS-PDU transfer, and to 0 for all other cells. AAL-5 detects the loss, misordering, or misinsertion of a cell by timeouts for reception, checks on the received length field, and the use of the CRC. One consequence of the AAL-5 CS-PDU structure is that it cannot interleave two or more CS-PDUs.

AAL-5 Multiplexing Example

Figure 10.17 shows an ATM terminal that has two inputs with two variable-length packets arriving simultaneously destined for a single ATM VC using the AAL-5 protocol. The network is not congested.

Video Transport Over ATM

Video is a part of both residential and business applications. Examples of these applications are as follows:

1. Residential applications:
    - Video-on-demand;
    - Near-video-on-demand.

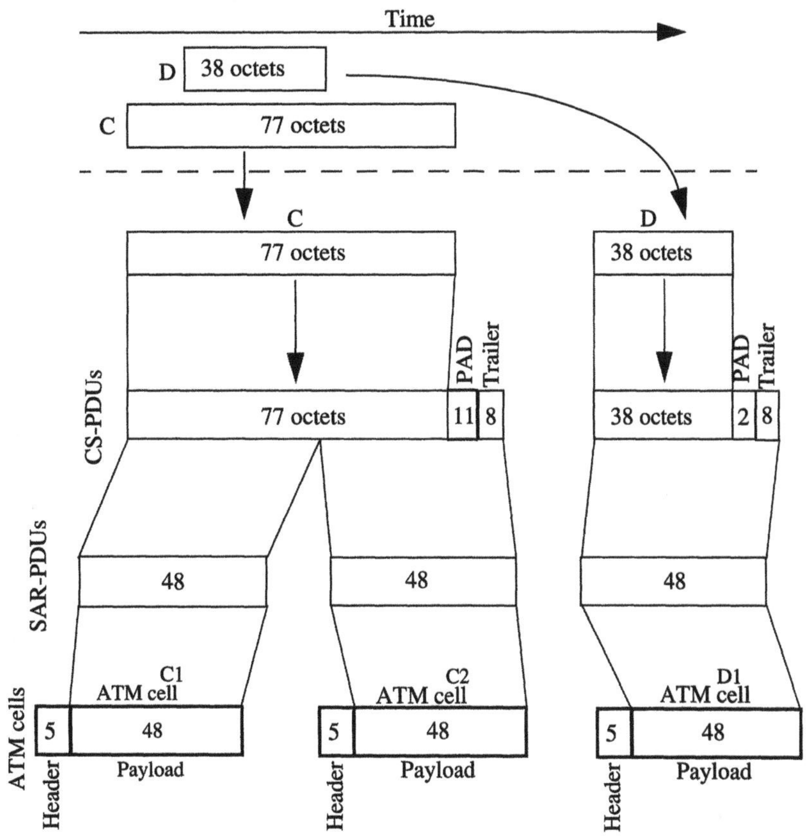

CS-PDUs

	C	D
PAD	11	2
Length	77	38

ATM cell transmission order

Transmission order	C1	C2	D1
PTI	000	001	001

FIGURE 10.17. AAL-5 multiplexing example.

2. Business applications:
    - Video conferencing;
    - Multimedia conferencing.

The transport of the video services can be accomplished over the ATM network platform very efficiently.

Residential Video Services

Video-on-demand– and near-video-on-demand–type services are mainly focused on the residential market.

Video-on-Demand

Video-on-demand (VoD) services are unidirectional and asymmetrical in nature. Digitally compressed and encoded video information is transported from a video server to a set-top terminal. The set-top terminal decodes and converts the digital streams back to the original analog signal, which is then passed to either a monitor or a television set for display. The set-top does not send any video information back to the video server. The control signals to set up the connection and further control of the audio/video can be sent via an out-of-band or in-band signaling channel to the network and to the video information provider (VIP).

Near-Video-on-Demand

Near-video-on-demand (NVoD) service does not require a dedicated connection to the VIP. The VIP simultaneously broadcasts different segments (staggercast) of a video program. The set-top (end user) sends control signals upstream to the network and to the VIP requesting a specific staggercast stream. This gives the end user to fast-forward or to rewind to a specific point of the video broadcast.

Video Service Components

The video services in general are shown in Figure 10.18.

Video Compression

The video information is coded in second Motion Pictures Experts Group (MPEG-2), which is the international standard for digitally compressed video, and then is transported over AAL-1 or AAL-5 as CBR service. Refer to ITU-T documents ISO/IEC 13818-1 (MPEG-2 System), 13818-2 (MPEG-2 Video), 13818-3 (MPEG-2 Audio), Bellcore document GR-2901-CORE, and ATM Forum document AF-SAA-0049.000 for more information.

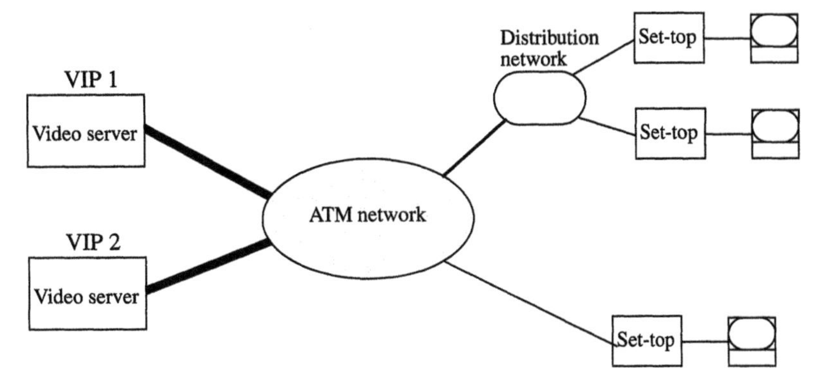

FIGURE 10.18. VoD services.

MPEG-2 is a superset of MPEG-1. The MPEG-1 system, video, and audio standards specify the coding of moving pictures and associated audio for digital storage media at up to about 1.5 Mbps. MPEG-1 is used for CD-ROM and Asynchronous Digital Subscriber Loop (ADSL).

MPEG-2 is standardized for transport of high-quality audio and video signals with higher bit rates (6 Mbps for CATV, 18 Mbps for HDTV). MPEG-2 applications are:

- Video-on-demand;
- Near-video-on-demand;
- Two-way communications;
- High definition TV (HDTV), 18Mbps;
- Cable;
- Satellite;
- Broadband networks.

The MPEG-2 deals with the multiplexing/demultiplexing and synchronization (clock recovery) of the audio, video, and user data of the program transmitted. MPEG-2 supports the following two types of system streams:

1. Program stream (PS);
2. Transport stream (TS).

Both the PS and the TS are made of elementary stream (ES) and packetized elementary stream (PES) as shown in Figure 10.19.

Protocol Reference Model

The video service protocol reference model is shown in Figure 10.20. Please refer to ATM Forum document AF-SAA-0049.000 for detail information.

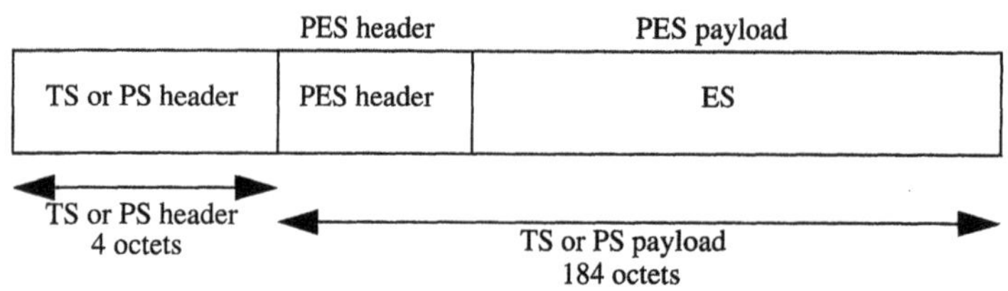

PS: program stream
TS: transport stream
PES: packetized elementary stream
ES: elementary stream

FIGURE 10.19. MPEG-2 program stream and transport stream.

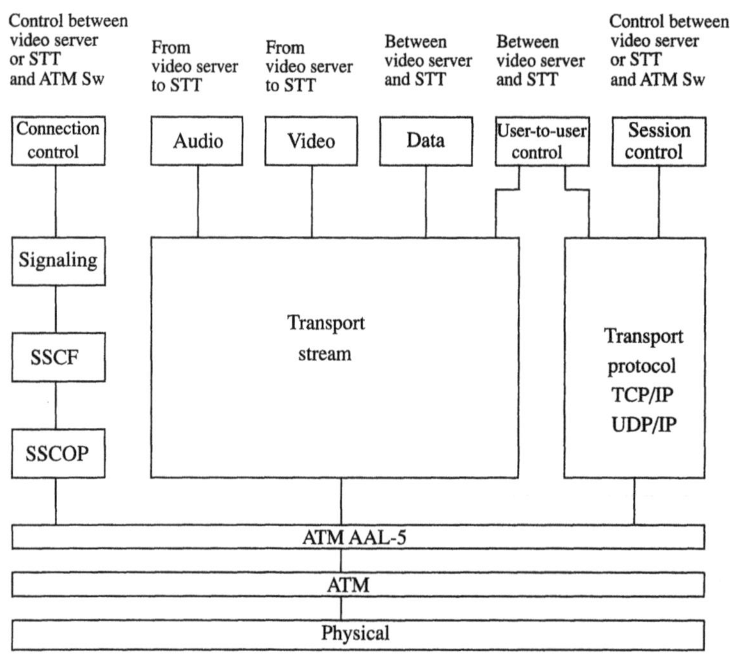

FIGURE 10.20. Video service protocol reference model.

Adaptation Process

The MPEG-2 single program transport stream (SPTS) packets are mapped into the ATM AAL-5. Normally, two SPTS are combined to create an AAL-5 CS-PDU. This is negotiated during the connection setup procedure for both SVC and PVC. The SAR process for MPEG-2 packets is shown in Figure 10.21.

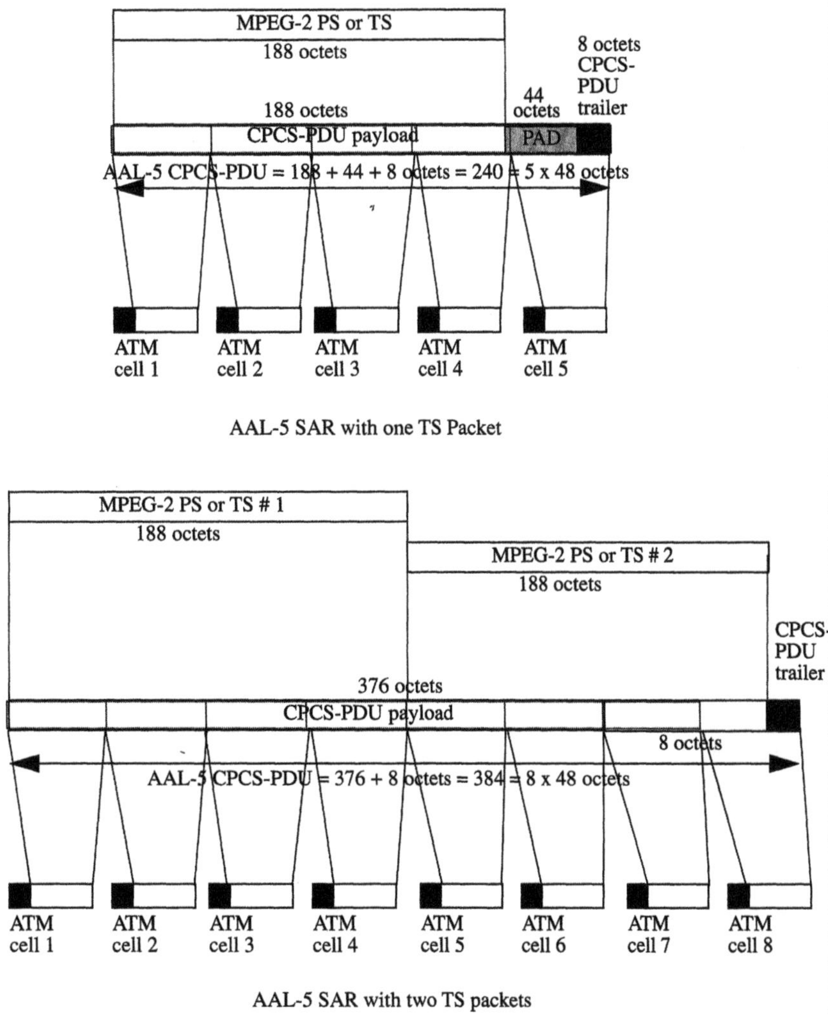

FIGURE 10.21. SAR process for two MPEG-2 packets.

Business Video Services

Video conferencing, distance learning, judicial conferencing, medical conferencing, and multimedia conferencing are mainly focused on the business market. These business applications require bidirectional and symmetric transport of MPEG-2 packets. Most of the above applications are based on interactive services. Refer to Bellcore document GR-1337-CORE for more information.

The business video services in general are shown in Figure 10.22.

Physical Layer

Video transport architecture is based on some of the following physical technologies:

- HFC (hybrid fiber-coax);
- FTTC (fiber to the curb);
- FTTH (fiber to the home);
- ADSL;
- SONET.

Multiprotocol Encapsulation Over ATM

Both local and wide area applications are required to use ATM-based network for global connectivity. The variable-length user information goes through AAL-5 SAR process. The short, fixed-length ATM cells are multiplexed over a single or multiple virtual connections.

The multiprotocol over AAL-5 is shown in the Figure 10.23.

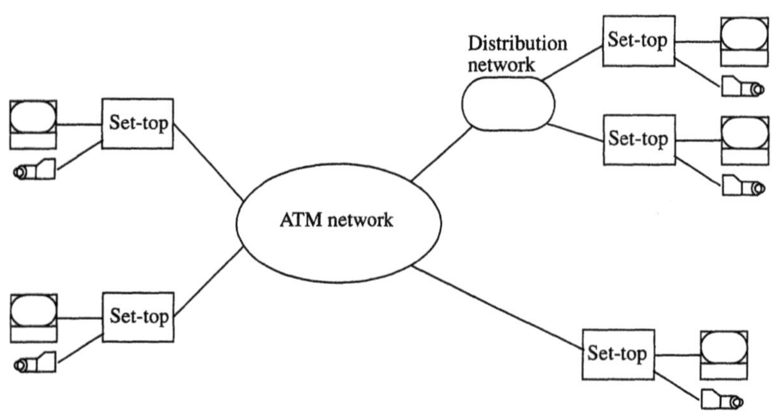

FIGURE 10.22. Business video services.

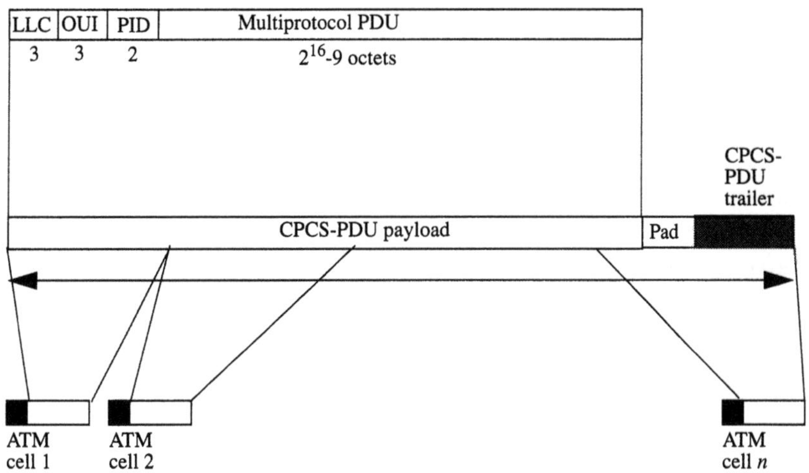

FIGURE 10.23. Multiprotocol over AAL-5.

The three-octet LLC encapsulation is required because several protocols are carried over the same virtual connection. The three-octet organizationally unique identifier (OUI) identifies an organization that administers the two-octet protocol identifier (PID). OUI and PID identify a unique routed or bridged protocol.

Refer to Internet Engineering Task Force (IETF) documents RFC 1483 for more information.

IP and ARP Over ATM

IP and ARP over ATM is intended for the ATM network to carry IP datagrams and ATM Address Resolution Protocol (ATMARP) requests and replies.

The internet was born when the U.S. Advanced Research Projects Agency network (ARPAnet) was split into a military network, and a public research network in 1983. The ARPAnet was introduced in early 1970s as a packet-switched network. It implemented Transmission Control Protocol/Internet Protocol (TCP/IP) in 1983 to replace the previous network protocols. The IETF was formed to define any future internet standards. Refer to IETF document RFC 1577 for more information about classical IP over ATM.

The IP reference model is shown in Figure 10.24.

Some of the applications supported by the transmission control protocol (TCP) include the following:

- The File Transfer Protocol (FTP) provides for file transfer, security log-in, and directory manipulation;
- The TELNET provides for remote terminal log-in.

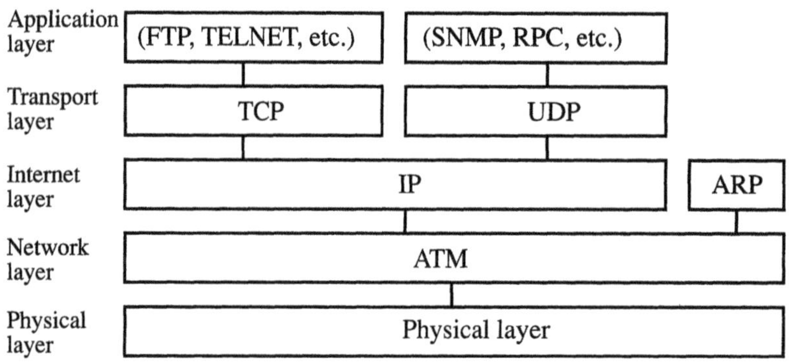

FIGURE 10.24. IP reference model.

Some of the applications supported by user datagram protocol (UDP) include the following:

- The Simple Network Management Protocol (SNMP) provides configuration setting, data retrieval, and alarms control;
- The Remote Procedure Call (RPC) allows application interaction over IP.

The IP datagram is shown in Figure 10.25. The IP version is defined by the version field. The header length is the length (32-bit word) of the IP header and the total length (TL) is the IP datagram length including the IP header. The 8-bit service type (ST) field specifies the reliability, throughput, and delay request. The segmentation and reassembly of IP datagrams are controlled by identification, flags, and fragment offset fields. The lifetime field specifies the time it can stay in the network before it is removed (killed) from the network. This field is decremented each time it passes through a node. The IP datagram carrying TCP or UDP packets is specified by the protocol field. The validity of the header field is protected by the header checksum field. The source IP address and the destination IP address specify 32-bit (2^{32} = 4 billion range) global network address.

The user data are placed in the data field. The Internet Assigned Numbers Authority (IANA) is responsible for IP network address assignment.

The E.164 public UNI addresses are analogous to the ethernet addresses. IP broadcast and multicast addressing are not supported in ATM.

The IP and ATMARP encapsulation is shown in Figure 10.26.

The presence of the SNMP header is indicated by LLC = AA AA 03 (Hex). The ethernet-type PID is indicated by OUI = 00 00 00 (Hex). The ethernet value of 08 00 (Hex) indicates that the PDU is classical IP.

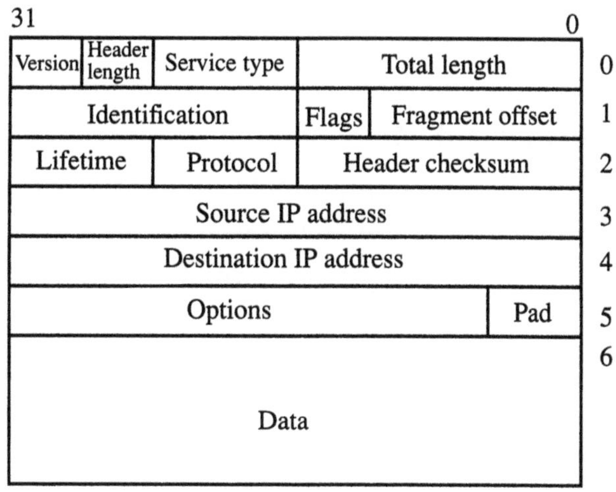

FIGURE 10.25. IP datagram.

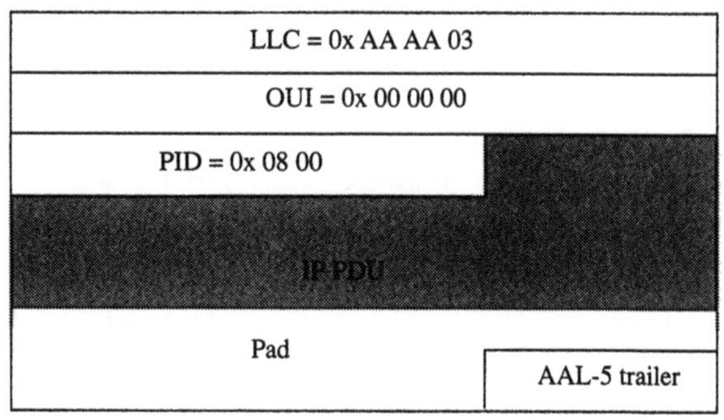

FIGURE 10.26. IP and ARP encapsulation.

IP Network

The IP network is made of numerous public and private networks. These networks are connected to a network access point (NAP) for interconnection. They also are interconnected by some peering arrangement. Each IP network is an FDDI ring or 100-Mbps ethernet. The multimedia terminals (e.g., PCs) are connected to the internet service providers (ISP). The ISPs are connected to any of the IP networks, or the NAP, or other larger ISPs. The ATM router converts IP packets into ATM cells and provides access to the ATM word. The ATM network is mainly used for trunking purposes.

The IP network is shown in Figure 10.27.

The ISP block diagram is shown in Figure 10.28.

Frame Relay Over ATM

The frame relay (FR) is very similar to the X.25 protocol and evolved from ISDN link access protocol D-channel (LAP-D) framing structure.

Refer to ATM Forum B-ICI specification (AF-BICI-0013.003) and ITU-T recommendation I.555 for more information.

The Q.922 FR frame structure is shown in Figure 10.29.

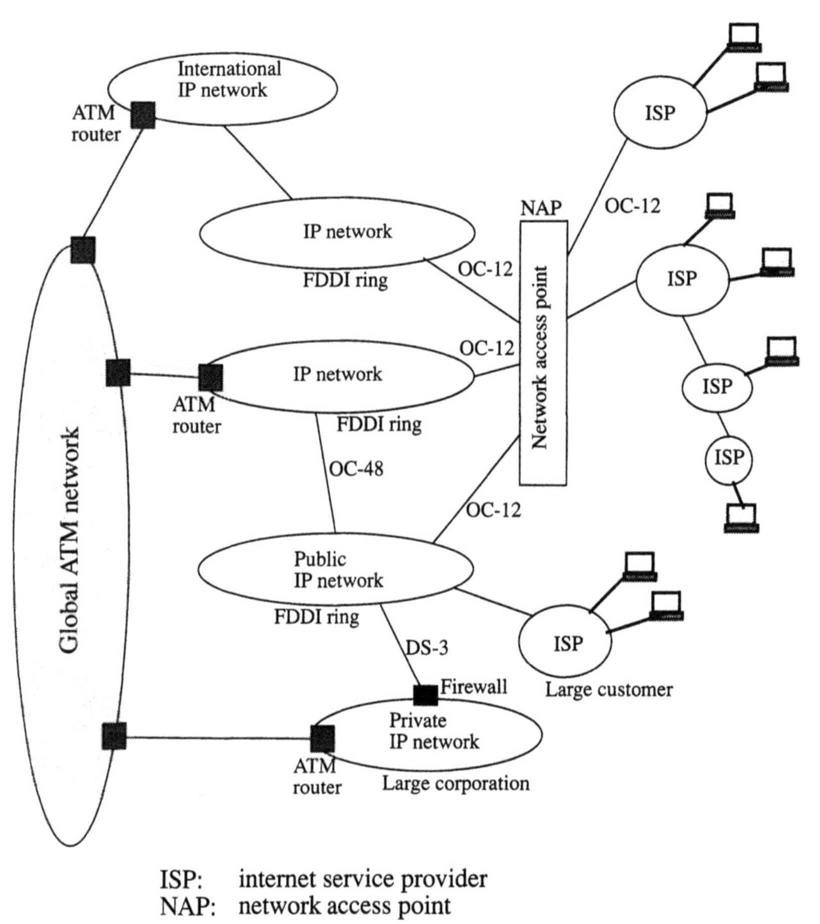

FIGURE 10.27. Internet protocol network.

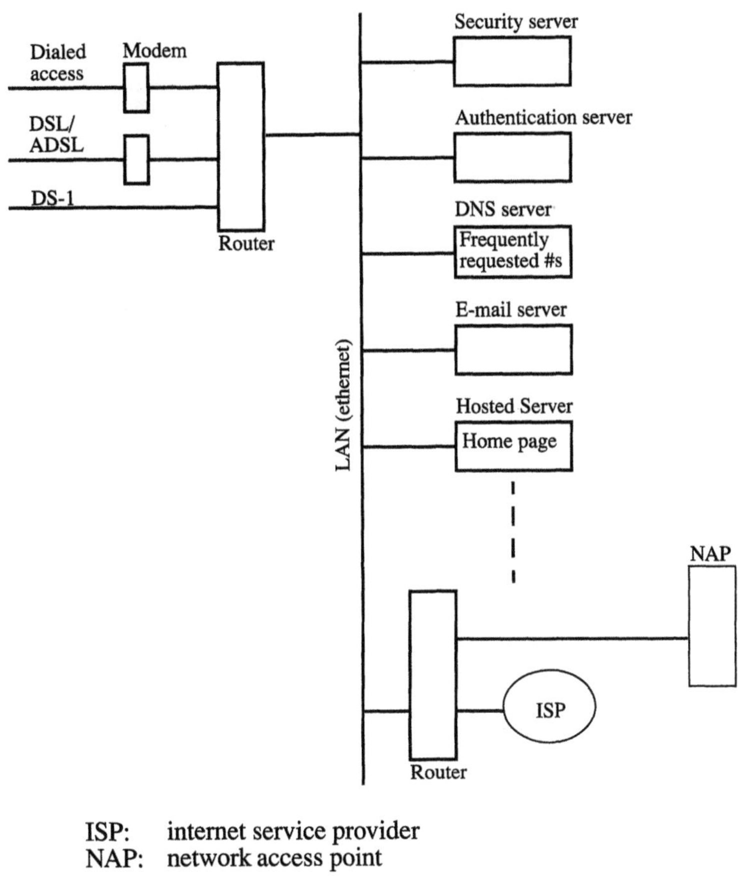

FIGURE 10.28. ISP block diagram.

FR frames are encapsulated by flags. The 10-bit DLCI can identify up to 1,024 unique virtual circuits per link. The same DLCI can be used in different links; therefore, it only has local meaning, like ATM VPI/VCI value. This is the limitation of the FR network, which is, fewer than 1,024 nodes. The network congestion either in the ingress or in the egress direction is expressed by setting the forward explicit congestion notification or the backward explicit congestion notification, respectively. The discard eligibility (DE) bit is set to indicate low-priority frame. Therefore, the frame with the DE set to one is discarded during the congestion state. Extended address (EA) bits are normally set to 0 and 1 as shown in the previous figure. These bits are used to extend the DLCI addressing range to three- and four-octet format.

The FR and ATM network interworking is shown in the Figure 10.30.

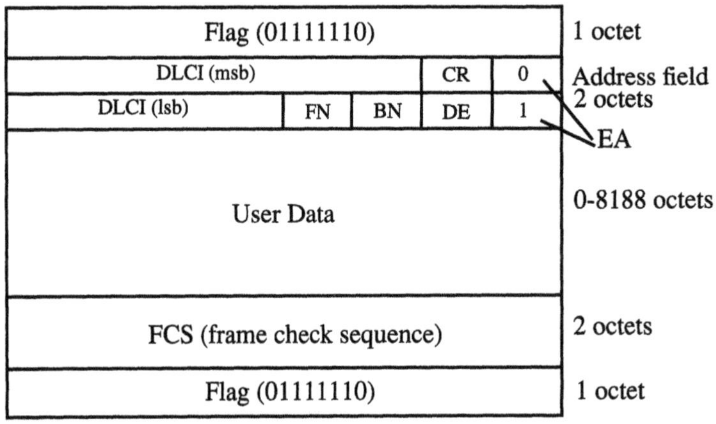

DLCI: data link connection identification
CR: command respond indicator
FN: forward explicit congestion notification
BN: backward explicit congestion notification
DE: discard eligibility
EA: extended address

FIGURE 10.29. Q.922 FR frame structure.

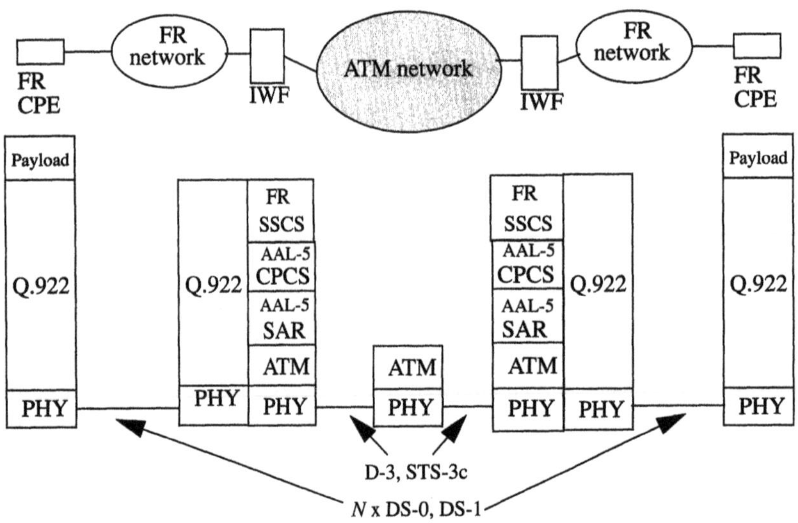

IWF: interworking function
FR: frame relay
SSCS: service-specific convergence sublayer

FIGURE 10.30. FR and ATM network interworking.

Switched Multimegabit Data Service Over ATM

SMDS was introduced by Bellcore for metropolitan area network (MAN). It supports connectionless integrated data, voice, and video services. ATM AAL-3/4 SAR process is used to transport SMDS layer 3 PDU across ATM network.

Refer to IEEE 802.6 for more information on SMDS protocol. The SMDS layer 3 PDU structure is shown in Figure 10.31.

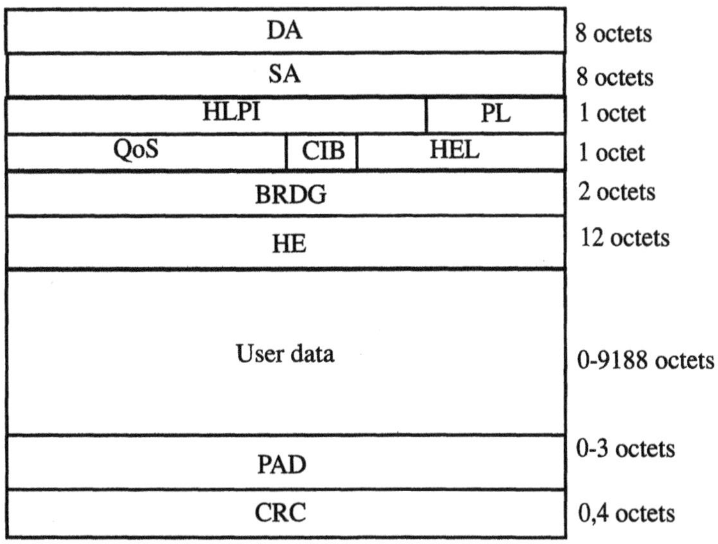

DA: destination address
SA: source address
HLPI: higher layer protocol indicator
PL: pad length
QoS: quality of service
CIB: CRC indicator bit
HEL: header extension length
BRDG: bridging
HE: header extension

FIGURE 10.31. SMDS layer 3 PDU structure.

IBM SNA Over ATM

IBM 3270 terminals can be distributed over the ATM network. They do not have to be collocated with the cluster controller. The SNA over ATM is shown in Figure 10.32.

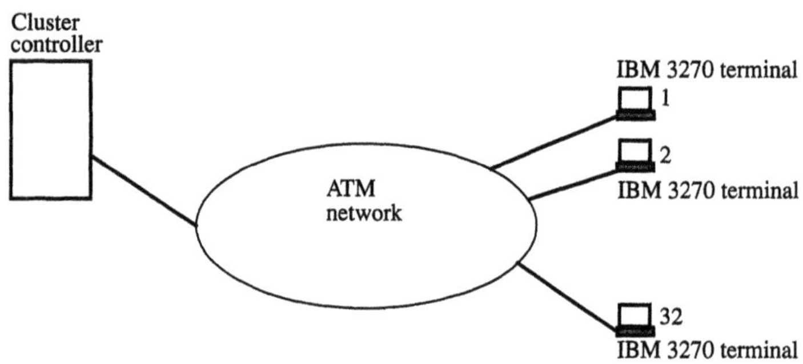

FIGURE 10.32. SNA over ATM.

CHAPTER 11

ATM Switching Matrix

ATM Switch

ATM has unique capabilities in the switching area. It offers a number of possibilities for switching that are not available in non-ATM systems. This is due to the segmentation of the streams into fixed-size cells. An ATM network is a system of interconnected ATM switches. An ATM switch is a set of input and output links across which ATM cells are sent and received. Figure 11.1 shows the basic ATM switching elements. A switching element is the basic unit of a switch fabric. The incoming cells are analyzed at the input ports and directed to the appropriate output ports. Basically, a switching element consists of an input controller for each input port and an output controller for each output port and an interconnection network.

The ATM switch basically has two functions as follows:

1. VPI/VCI translation;
2. Cell transport from its input to its output ports.

First, it interprets the VPI/VCI field values in the incoming ATM cell header. Second, it routes the ATM cell to the appropriate output port or ports.

An ATM switch should be designed with consideration of the following characteristics:

- Cell throughput should be in Gbps range;
- Cross-node delay should be minimum;
- Cell loss should be minimum.

There are six general types of switching fabrics used in ATM switching. These are the following:

- Matrix switch;
- Shared medium (bus) switch;

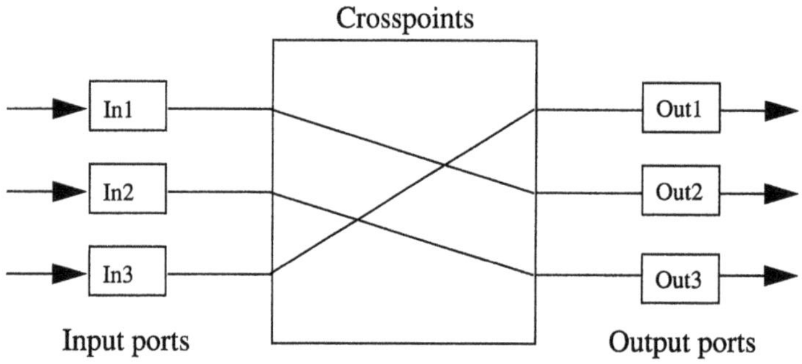

FIGURE 11.1. Basic ATM switching elements.

- Shared memory switch;
- Self-routing switch;
- Table-based switch;
- Backplane switch.

ATM Matrix Switch

ATM switching is generally regarded as the same as the various forms of matrix switches. A comparison may be drawn between the matrix switch and a version of the cross-bar type switches used for switching telephone calls, but much faster than the latter. The matrix switch uses a matrix type set of internal switching elements through which ATM cells are routed from their input ports to their output ports based on their VPI/VCI fields. Sometimes these are referred to as M x N switches where the M stands for the number of input ports and the N stands for the number of output ports. Figure 11.2 shows the basic matrix switch.

A drawback of the matrix switch, presently, is the need for buffering internally to prevent congestion at the output ports. Another is that most switch fabrics do not themselves support multicasting. This requires a separate copy fabric appended in some way. There are three types of matrix switches:

1. Input buffer matrix switch;
2. Output buffer matrix switch;
3. Crosspoint buffer matrix switch.

Input Buffer Matrix Switch

The input buffer matrix switch (IBMS) buffers are placed at the input ports of the switch. When two or more ATM cells are destined for the same output port simultaneously, one of the cells must be buffered at the input port. If it is not buffered, then one of the cells will be blocked, which is generally not acceptable. Cells following the blocked head of the queue cell are also blocked; even they are destined for another available port. Figure 11.3 shows the IBMS.

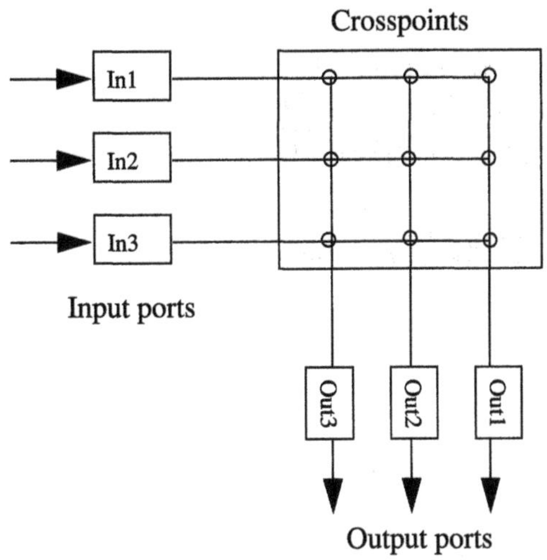

FIGURE 11.2. Basic matrix switch.

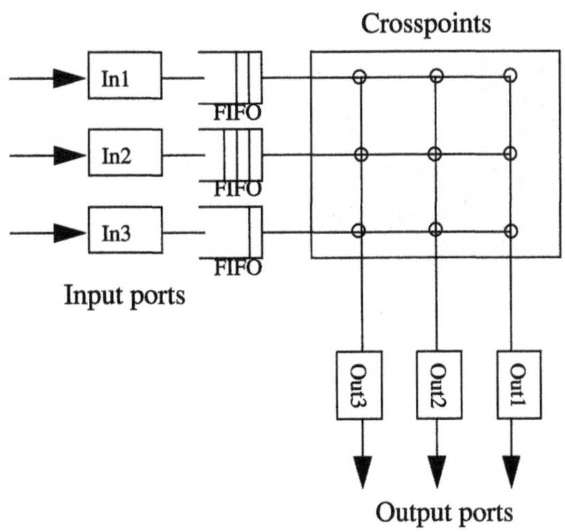

FIGURE 11.3. Input buffer matrix switch.

Output Buffer Matrix Switch

The output buffer matrix switch (OBMS) operates similarly to the IBMS except that the buffers are located at the output ports. When two or more cells from the input ports are delivered to the same output port, instead of discarding a cell, it is buffered until it can be read by the corresponding output port. Figure 11.4 shows the OBMS.

An internal nonblocking OBMS can still block cells at the output ports during contention. OBMS has better switch throughput than IBMS, because in IBMS only one of the contending cells from different input ports can be delivered to the output port.

Crosspoint Buffer Matrix Switch

The crosspoint buffer matrix switch (CBMS) has buffers in every crosspoint. Cells are buffered at the crosspoints when conflict occurs. If there are cells in more than one buffer destined for the same output, an arbitration strategy is implemented to decide which buffer will be served first. Figure 11.5 shows the CBMS.

In CBMS, crosspoint buffers eliminate the head of the queue cell blocking, unlike it is in IBMS. CBMS does not require the additional connectivity complexity introduced by OBMS. Internal cell blocking probability is reduced because of the crosspoint buffers. But, large buffers are required since cells may arrive at the crosspoint from all input ports. Large buffers also introduce undesired cell delay variation in the network.

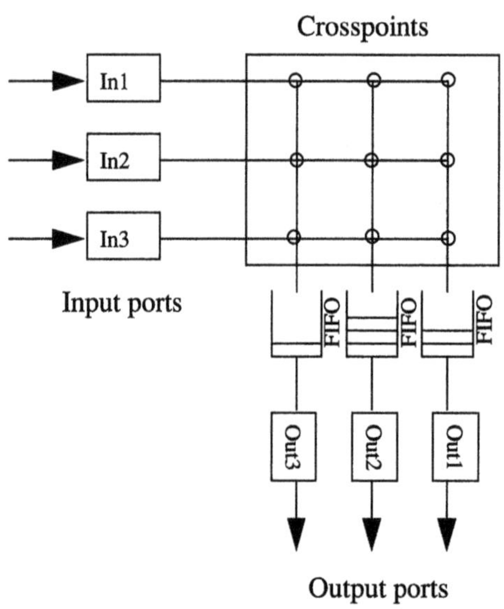

FIGURE 11.4. Output buffer matrix switch.

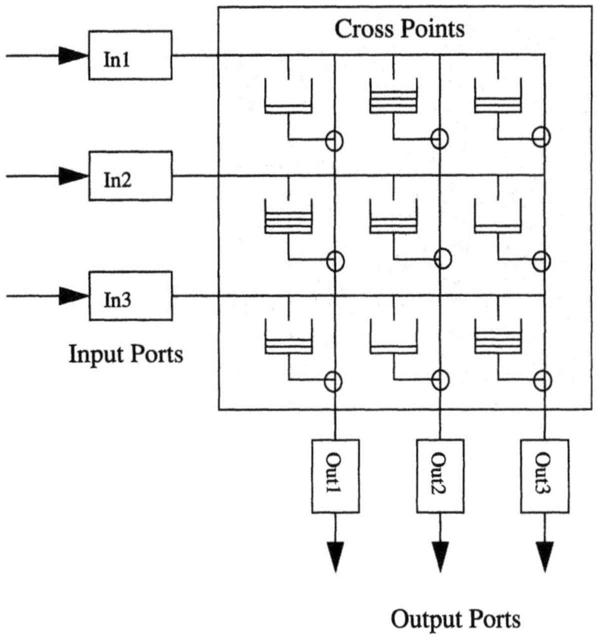

FIGURE 11.5. Cross point buffer matrix switch.

Arbitration Strategies

Arbitration strategies are required if more than one cell compete simultaneously for the same output. Only one cell will be served, and all remaining cells are delayed for their turns. To decide the winning cell, the following objectives are kept in mind:

- All queues should be given a fair chance for servicing cells;
- Cell loss should be kept at a minimum;
- Cells should be serviced at a constant rate from a given queue with a minimal cell delay variance.

The following arbitration strategies are most commonly used in the telecommunications system:

- Random;
- Cyclic;
- State dependent;
- Delay dependent.

Random

The servicing buffer (line) is chosen randomly from all competing buffer (line) to determine the winning cell. This is very easy to implement.

Cyclic

The buffers (lines) are served in a cycle to determine the winning cell. A shift register or counter can be used to achieve this function. This is very easy to implement. It works better when all the input ports carry the same amount of traffic load. Head-of-the-queue blocking may occur because of contention. An extra circuit is required to skip an input port when it has nothing to send. See Figure 11.6 for a random and cyclic arbitration block diagram.

State Dependent

The first cell (head of the cell) from the longest queue is served first to determine the winning cell. The mechanism to determine the number of cells in each queue (buffer) is implemented in this strategy. It is more complex to implement because the fill level of all the queues has to be monitored. It is better when traffic distribution is uneven among the input ports. See Figure 11.7 for a state-dependent arbitration block diagram.

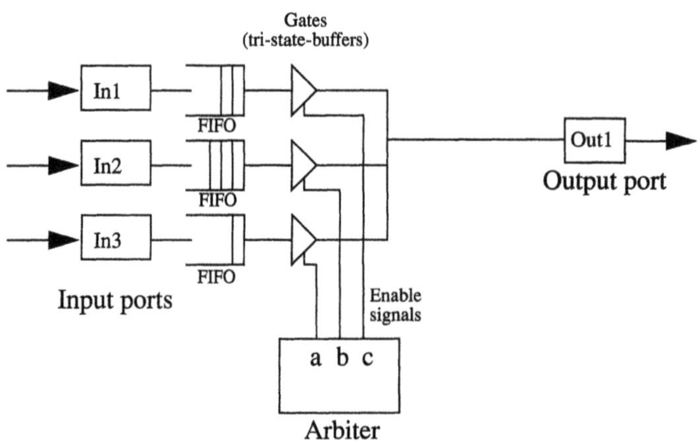

Random: enable signals a, b, c are asserted randomly
Cyclic: enable signals a, b, c are asserted cyclicly

FIGURE 11.6. Arbitration (random and cyclic).

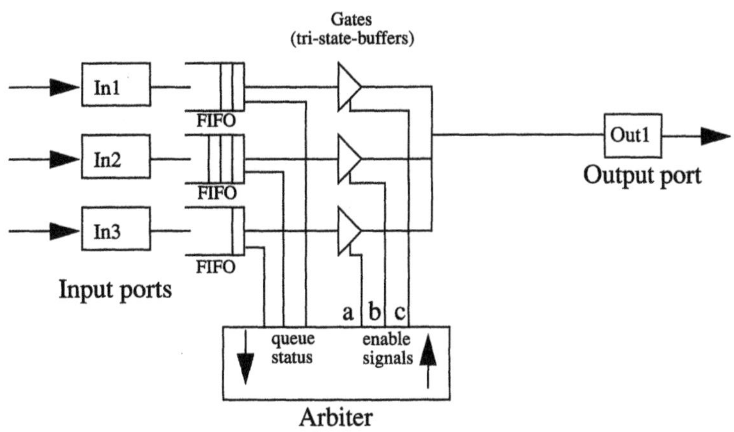

Enable signals are asserted in the following order: b, c, and a.

FIGURE 11.7. Arbitration (state dependent).

Delay Dependent

The servicing buffer is a global FIFO. All cells destined for a single output are fed into this FIFO. This global FIFO is served one cell at a time. See Figure 11.8 for a delay-dependent arbitration block diagram.

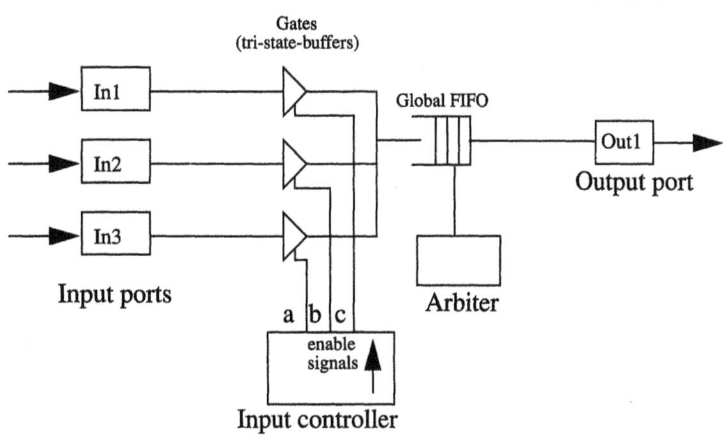

Enable signals a, b, c are asserted when cells arrive at input ports.

FIGURE 11.8. Arbitration (delay-dependent).

ATM Shared Medium (Bus) Switch

The traditional method in the computer industry for moving data within a machine to a variety of destinations is the bus. This is a unifying characteristic shared by everything from microprocessors to mainframes. Operating as a cell switch, the bus-based system processes the arriving cell in an incoming line card that examines the header and forwards the cell over the bus to the proper outgoing line card. The outgoing section of all line cards then examines the cell header to determine whether the cell is routed for it. If so, it extracts the cell and stores the cell in its buffer. Figure 11.9 shows the shared bus switch.

With a bus-based switch, the overhead allocation is kept small by a round-robin scheduling queue or by request/grant cycles. Because the fact that the cell size is small and fixed, delays can be held to a low level.

In a shared bus switch, all ATM cells are transferred over a single bus. Only one input port at a time can transfer an ATM cell over the bus to the output port or ports. All of the output ports monitor the bus constantly. If the ATM cell address matches with the output port address, the cell is then captured by that output port only. The rest of the output ports will continue to monitor for cells with a corresponding address in the next cell slot. This bus is time-division-multiplexed, which means each input port has its own timeslot to transfer cells. Buffers are used to store ATM cells as the need occurs. Normally, output buffers are used to store ATM cells at the output ports. The aggregate bandwidth of all of the input ports is equal to the bus bandwidth. Therefore, the input ports cannot overload the bus. When two or more cells arrive at the same output port simultaneously, cells are buffered at that point.

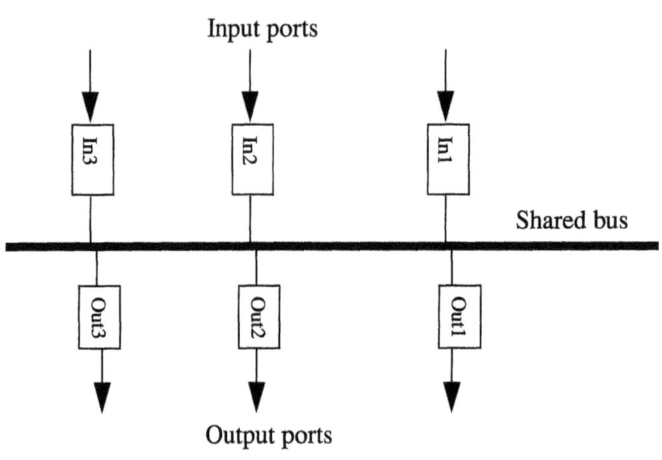

Note: Maximal bus speed cannot exceed the combined speed of all the input ports.

FIGURE 11.9. Shared bus switch.

Sometimes multiple buffers are used for an output port. High-priority cells are stored in one buffer and the lower priority cells are stored in another buffer.

Buffers (random access memory) are expensive; therefore, the size of the buffers are kept to a minimum. Cell loss occurs when buffers overflow. To avoid cell loss, several threshold levels are set for each output buffer. When these threshold points are approached, back pressure signals are generated to all the input ports, so that all the input ports can stop transferring ATM cells to that particular output port whose buffers are near overflow condition. The only way an input port can stop the flow of ATM cells without discarding is by buffering.

The schedule by which each input port transfers its cells is normally a round-robin. An arbiter can be used to schedule a more complex port service mechanism; that is when an input port has no cells to transfer, the input port can be skipped.

Priorities may be established for certain input ports. The ideal arbiter function is to monitor back pressure signals from all the output ports.

It should react in such a fashion that only the ATM cells designated for the output port, this is indicating a near full buffer condition, are held in a separate buffer at the input port. This allows the transfer of subsequently arriving cells that are destined for other output ports and prevents stopping the pipeline for the duration of the active back pressure signal. Figure 11.10 shows the shared bus switch with back pressure mechanism.

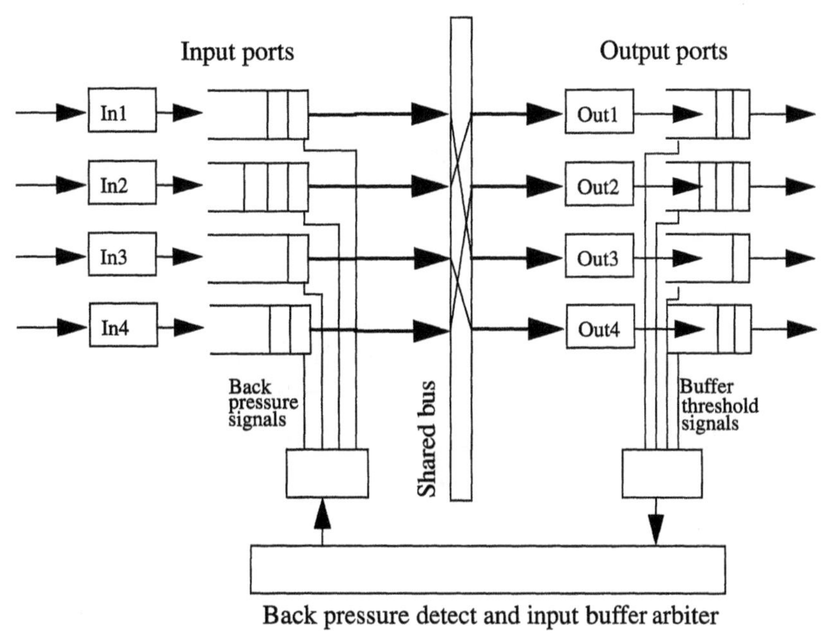

FIGURE 11.10. Shared bus switch with back pressure mechanism.

The bus-based ATM switches are relatively simple to implement and can easily support multicasting. As long as the speed of the bus interface logic are within the range of the CMOS technology, the most economical system to build would be the shared-bus system. With a 64-bit wide bus, the result is a maximal speed of 3.2 Gbps, assuming a 50-MHz system.

ATM Shared Memory Switch

The shared-memory architecture is based on a very large and fast memory in which cells are deposited by the input port logic. Meanwhile, a chain of pointers is built to indicate where the cells for a particular output port are to be located. These cells are then retrieved from the memory by the output circuitry and follow the pointers to the proper output port. The shared-memory switch is shown in Figure 11.11.

Higher speeds can be attained with the shared-memory system than with the shared-bus system. The data can be multiplexed up to significant speeds with parallel access to many bit lines, perhaps 256 or more, to reach tens of gigabits. For simplicity, the

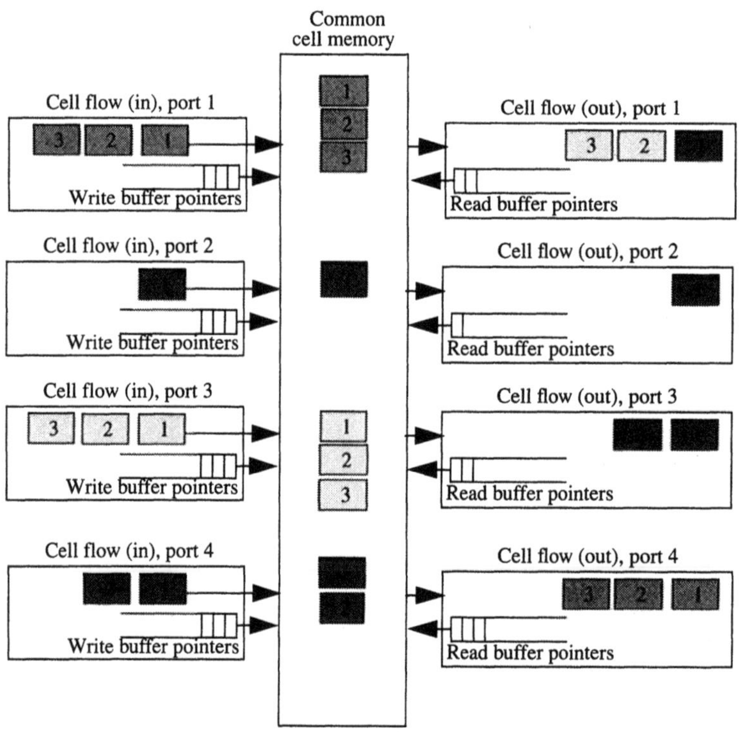

FIGURE 11.11. Shared-memory switch.

multiplexing steps are handled synchronously even at the loss of some capacity. Even if the input and output ports are somewhat limited, this is a powerful approach.

ATM Self-Routing Switch

The self-routing switch has been a remarkable development. In this design, the cell header carries the information that is interpreted by the switch fabric to route the cell through the various gates to its destination. The most common of these is the *banyan switch.* The banyan switch was named after the tropical tree with its complex configuration of trunk and branches. An advantage of this type of switch is that it is able to provide a high degree of parallelism, having for practical purposes, a miniprocessor at each switch point. Figure 11.12 shows the self-routing switch operation.

An analogy may be drawn between this switch and a railway switching yard. The "train" of bytes follows the "engine," the header cell, as it goes through the fabric. As the cell travels through, each switch point routes the cell based on the first bit of the header. This process consumes the first bit and the next switch point uses the new first bit to perform the same routing process. Because this design is highly regular, it is very capable of being laid down on silicon.

Although banyan switches are capable of large capacity, they are complex and present some designing challenges still. For instance, the design of the buffers at the switch point to deal with collisions that may occur when two cells arrive at the same time at different input ports. Obviously, only one cell can be allowed to go through while the other must wait, and while waiting it must not get out of sequence in respect to other cells that arrived at the same input port.

Another difficulty with banyan switches is the support of broadcast and multicast transmissions. It is simple enough to implement on a bus-based system. Trying to

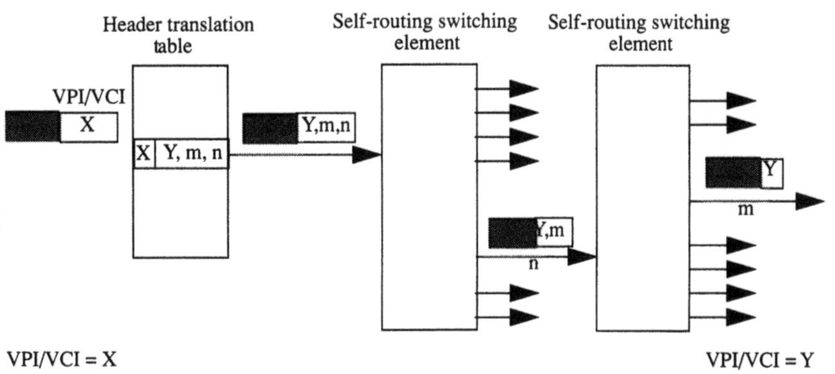

FIGURE 11.12. Self-routing switch operation.

use a banyan switch for the same transmissions would require a cell replication capability as complex as the core of the banyan switch itself.

ATM Table-Based Switch

A table-based switch is similar to a self-routing switch, where the ATM cell header (VPI/VCI fields) are used to look up tables at each switching point. A new VPC/VCI and an output port number are retrieved from the table. The ATM cell with a new VPI/VCI number is transferred to the output port, which was retrieved from the table. The same action is taken in the next switching element. Figure 11.13 shows the table-based switch operation.

ATM Backplane Switch

In the backplane system, ATM cells are transported across a bus that links two ATM modules. The modules receive cells based on their VPI/VPC values. Because the backplane design is well understood, ATM switches based on the backplane design are less expensive to develop. The design also lends itself well to supporting multicasting, which is essential for LANs.

An advantage of backplane design is the ability to integrate with existing network systems. This allows an economy of design, where, for instance, multiple modules share the bandwidth of a high-speed bus because each module by itself may not need the entire bandwidth of an ATM link.

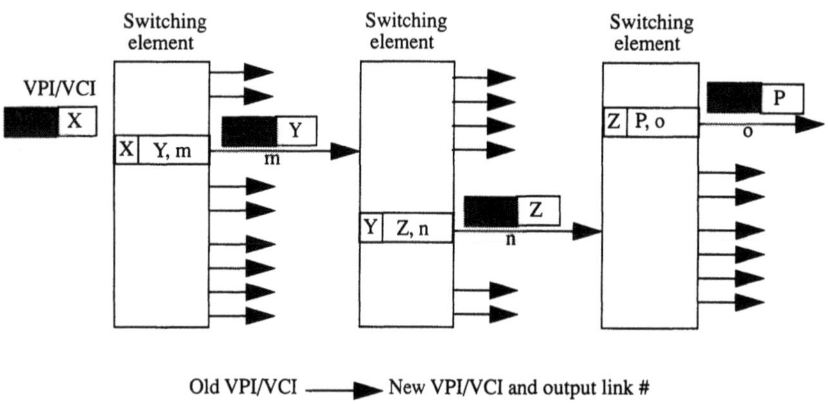

FIGURE 11.13. Table-based switch operation.

Operating at gigabit speed over a high speed, a backplane can give a module a greater peak data rate than could a single ATM link on an ATM switch, which is typically restricted to 100 to 155 Mbps. For this reason, almost all the early private ATM switches used ATM-base backplanes of different sorts. The recent development of ATM backplanes has been a result of trying to solve the bandwidth problem for today's LANs.

The real switch is the backplane bus, which allocates one cell at a time. For example, the buses used in LANS, such as Ethernet, are really the same as a computer's internal bus. The difference being that they are serialized and have an allocation system. Figure 11.14 shows the ATM backplane switch.

A distinction between the matrix switch and the current networking equipment is that a LAN, for instance, is a shared medium. In a shared medium, the bandwidth is shared by all users, with each user receiving only a fraction of the bandwidth. In the matrix switch, the bandwidth of each port is dedicated to a single user. Ideally, the aggregate bandwidth of the switch is the sum of that of each of the interfaces.

During the transition from backplane based switching to matrix switching, it is likely that designs will include ATM backplanes, because they will provide the smoothest transition for private ATM networks.

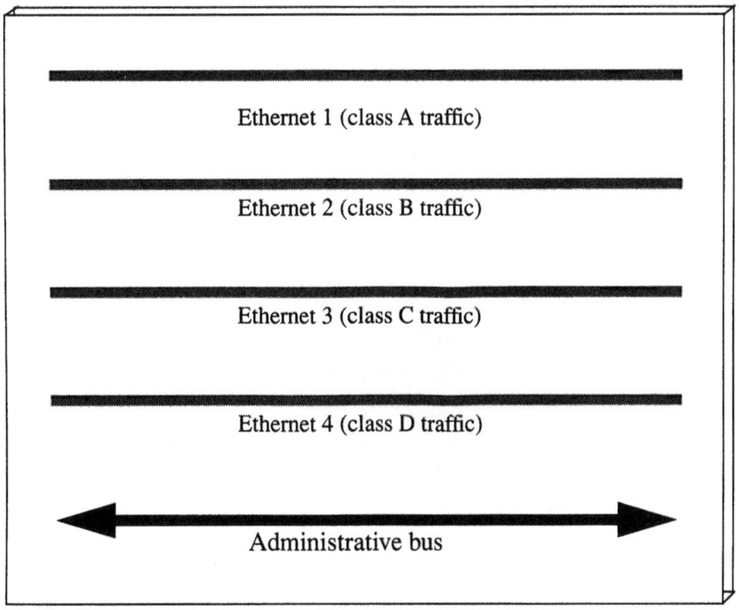

FIGURE 11.14. ATM backplane switch.

ATM Switching Networks

The switching networks can be divided into the following two stages:

1. Single-stage network;
2. Multi-stage network.

Single-Stage Network

A single-stage network is built up by a single stage of switching elements that are connected with the input and output ports of a switching network. The single-stage networks can be subdivided into the following types:

- Extended switching matrix;
- Funnel-type network;
- Shuffle exchange network.

Extended Switching Matrix

An extended switching matrix network is formed using basic *n* x *n* switching elements. These switching elements are extended by adding additional *n* inputs and *n* outputs. See Figure 11.15.

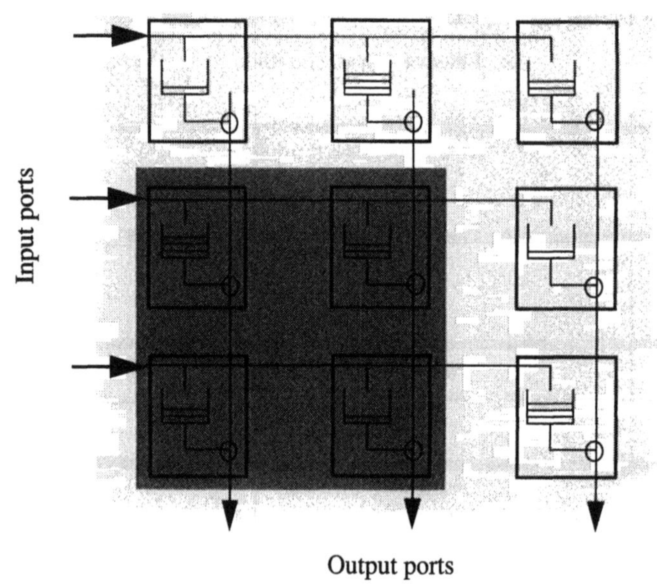

FIGURE 11.15. Extended switching matrix network.

Funnel-Type Network

In the funnel-type network, all the switching elements are interconnected in a funnel-like structure, as shown in Figure 11.16. All the switching elements contain $2n$ inputs and n outputs.

Shuffle Exchange Network

This network is also called a "recirculating network," because cells may pass through the network several times before they arrive at their proper output ports. See Figure 11.17.

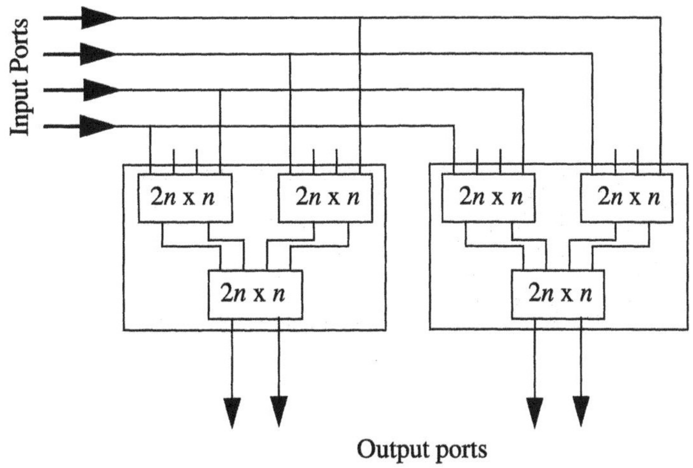

FIGURE 11.16. Funnel-type network.

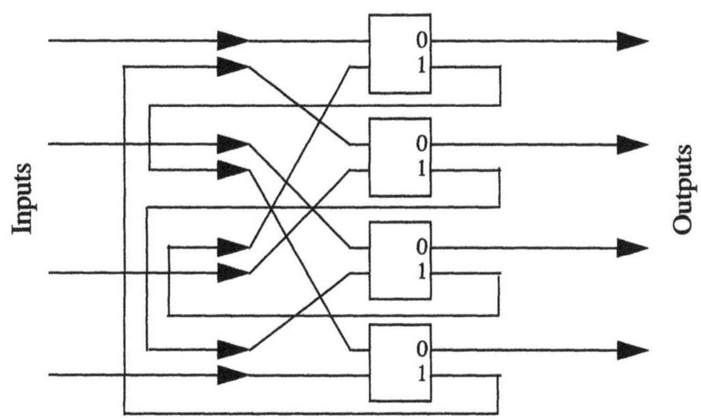

FIGURE 11.17. Shuffle exchange network.

Multistage Network

A multistage network is built up by multiple stages of switching elements. All stages are interconnected by a specific link pattern. The multistage networks can be subdivided into the following types:

- Single-path network;
- Multiple-path network.

Single-Path Network (Banyan Network)

In this type of network, there is only one path between a given input port and an output port. These networks are known as "banyan networks." The main advantage of these networks is the simple routing algorithm. The internal links are shared among other paths as well. This may cause internal blocking because of the internal links being used for other active connections. See Figure 11.18.

In the L-level banyan network, only the switching elements of the adjacent stages are interconnected. Each path passes through the L-stages. Delta networks fall into this category. The delta networks have the self-routing property that is independent of the input port at which the ATM cell enters the delta network: it will always arrive at the correct output.

These self-routing networks are suitable for packet-switching applications. By implementing routing function in hardware, high throughput can be obtained. Cells are also routed in parallel. Special provisions must be taken to avoid cell loss due to internal blocking. Several ways internal cell blocking can be reduced, such as the following:

- Implement buffers in every switching element within the switching fabric;
- Faster internal link speed compared with external speed;
- Implement back-pressure mechanism;
- Provide multiple paths using multiple network.

These networks are further subdivided into two types of network listed below:

1. Regular banyan;
2. Irregular banyan.

Regular Banyan

Regular banyan networks are made of identical switching elements.

Irregular Banyan

Irregular banyan networks are made of different types of switching elements.

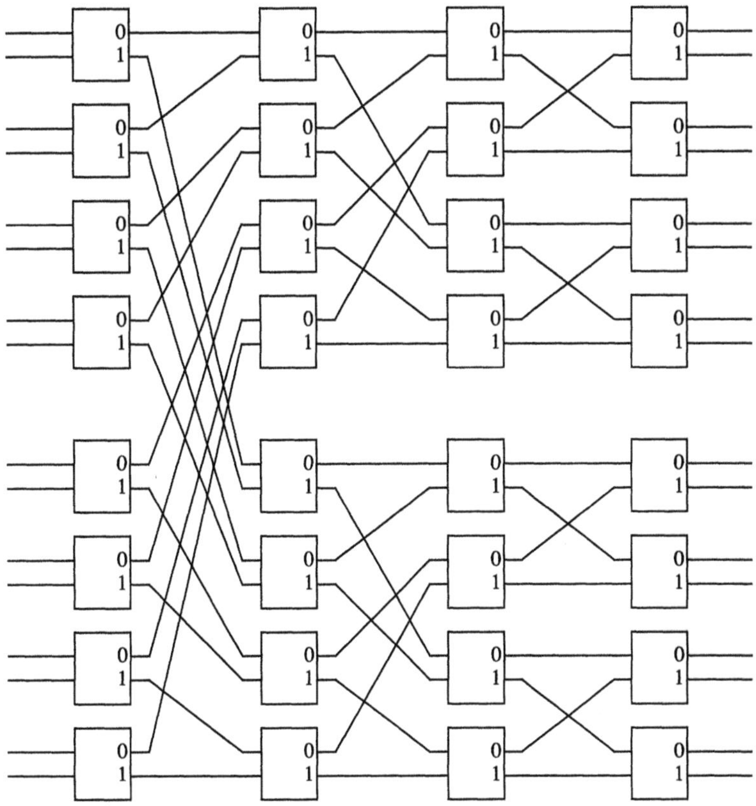

FIGURE 11.18. Single-path network.

Multiple-Path Network

In this type of network, there is more than one path between a given input port and an output port. The blocking of cells is minimized because of alternating paths between input and output ports. The internal paths are commonly determined during initialization phase. The multiple-path networks are subdivided into the following two categories:

1. Folded network;
2. Unfolded network.

Folded Network

These networks use bidirectional links, and all input ports and all output ports are located at the same side of the switching network.

If an input port and an output port are connected to the same switching element, cells can be routed in that switching element without having to pass them through the last stage. The number of stages required by these types of networks is determined by the locations of their input and output ports. See Figure 11.19.

Unfolded Network

These networks use unidirectional links. All input ports are in one side of the switching network, and all output ports are located at the other side of the switching network.

All cells are routed through all the stages of the switching network.

See Figure 11.20.

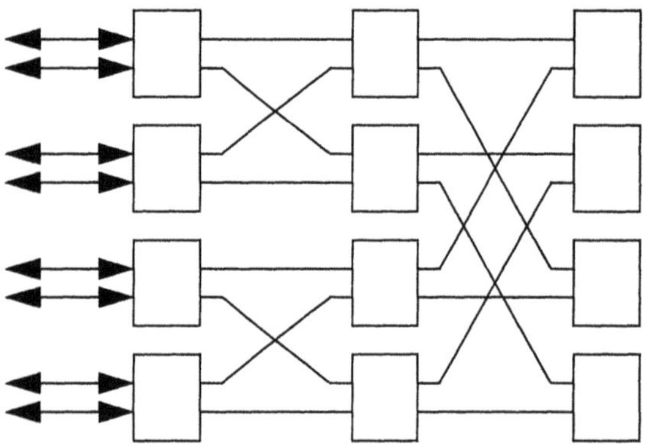

FIGURE 11.19. Three-stage folded network.

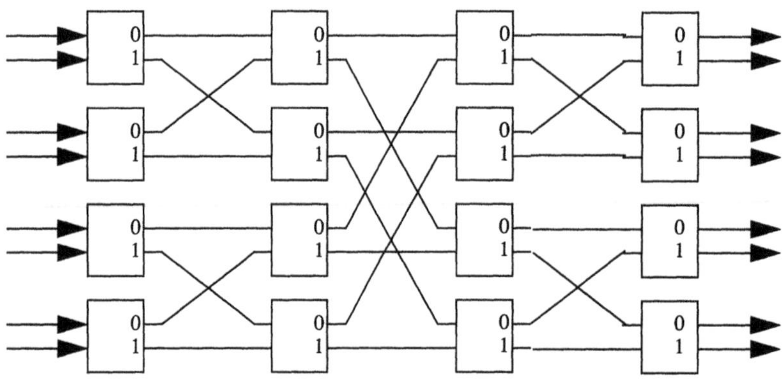

FIGURE 11.20. Four-stage unfolded network.

Examples of Combinational Network

A distribution network can be placed in front of a banyan network to distribute ATM cells evenly over all inputs of the banyan network. Cell blocking is minimized at the expense of cell sequence integrity of a connection. Therefore, additional hardware is required at the output for resequencing of cells. See Figure 11.21.

A combination of a sorting network and a trap network placed in front of a banyan network is shown in Figure 11.22. The sorting network rearranges the ATM cells according to their destination and passes them to the trap network. The function of the trap network is to identify all the cells that are bound for a particular output. It only passes one cell to the banyan network, which is bound for that particular output. The remaining trapped cells are time-stamped and routed back to the sorting

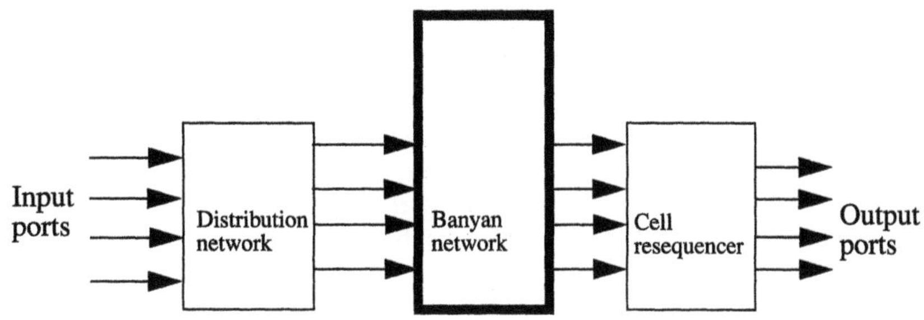

FIGURE 11.21. Distribution/banyan network.

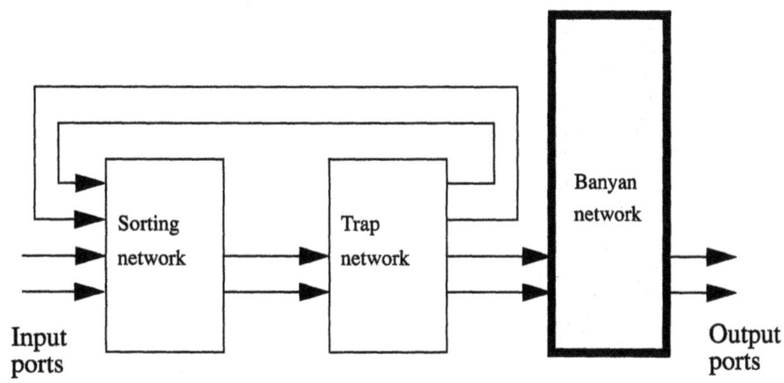

FIGURE 11.22. Sorting/trap/banyan network.

network. The sorting network assigns a higher priority to these cells to maintain cell-sequence integrity. If a particular cell is trapped *N* number of times (*N* is set by software as a cell lifetime), it is discarded from the network.

Examples of Delta Network

In this example, the basic switching blocks of 2 x 2 are organized in four stages, each stage having eight elements. The interconnection of the stages is such that all 16 input ports of the fabric can reach all 16 output ports, which is required for full interconnectivity.

A single string of digits is used as routing tag, containing the address of the outlet. The routing tag is used to route cells in each stage of the switching fabric. If the bit is a 0, the upper outlet of the 2 x 2 switching element is selected, regardless of the inlet at which the cell arrived.

In Figure 11.23 it is shown that a cell arrives at output port "0" which originates from input port "b" with a routing tag 1110. At every stage, the same decision (selection of upper or lower outlet) is made. The routing tag is shifted internally over one bit position, to allow each switching element always to interpret the first bit. This example shows switching of cells without any collision.

The basic self-routing switching block can be built using larger switching elements; it is not restricted to the 2 x 2 element. In that case, however, the routing function will require more bits. For instance, with 32 x 32 switching elements, a 5-bit routing tag is required to identify outlet in each stage.

Figure 11.24 shows how collisions may occur in the same network. Cells from input ports "a" and "n" arrive at the output port "c." Collision occurs at the forth stage. Similarly, cells from input ports "i" and "o" are bound for output ports "f" and "e." Collision occurs at the third stage.

4 x 4 ATM Switch Design Example

Switch

A 4 x 4 ATM switch based on an 8-bit sliced shared bus is described in this example. It provides switching functions on a per-cell basis. The structure of the switch is shown in Figure 11.25. The switch fabric consists of a single-stage bus switching. It is an output buffer switch with a buffer for each output port. Each of the four buffers in the switch can store up to 16 cells. Each cell, from one of the four independent incoming lines, is output to a desired line according to the header of the cell.

Figure 11.26 shows the switch operations (point-to-point connection). The cell-switching connection is controlled by referring to the ADDRESS (16-bit) field in

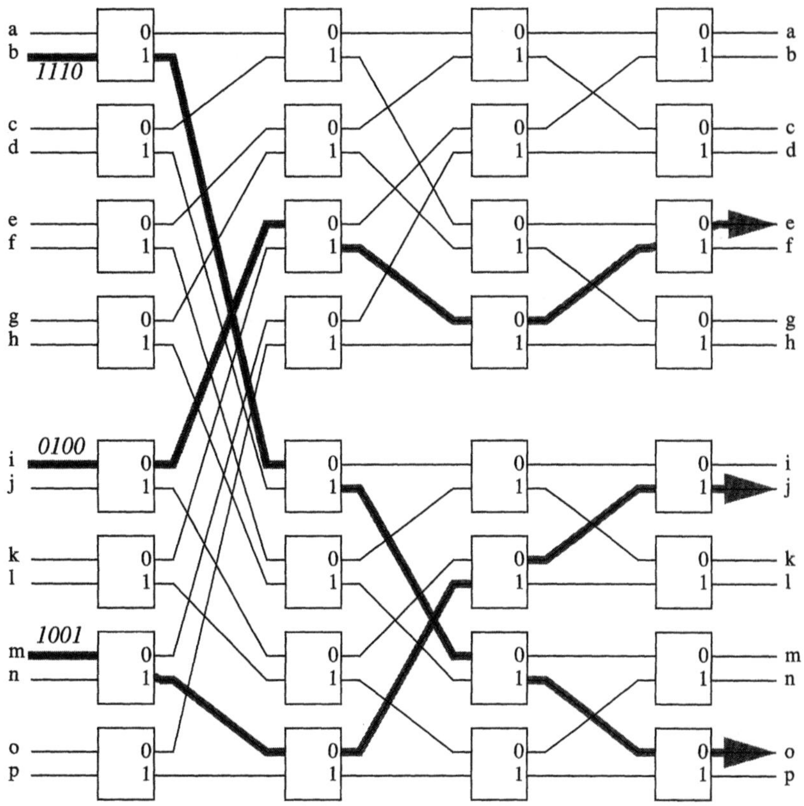

FIGURE 11.23. Four-stage delta network (no collision).

the header of a cell. Each cell arriving at the ATM switch has switch overhead information, which is appended to the cell by a previous process in the line card. The PORT data field in the ADDRESS of the switch overhead represents the destination port address of the cell. All cells arriving at the input ports are multiplexed on the time-division bus contained within the 4 x 4 ATM switch. A FIFO buffer with an address filter is located at the stage after the time multiplexing bus. A buffer and address filter are associated with each output port. Of the cells arriving from the time-division bus, an address filter writes to the FIFO buffer only those cells whose PORT indicates its own port number. The cells addressed to the address filter's port and written to the buffer are then read out from the buffer at the output port's speed. The 4 x 4 ATM switch only refers to the PORT in the header and switches the cells; it does not revise any information in the cell headers or cell payload (all information in the cells are preserved).

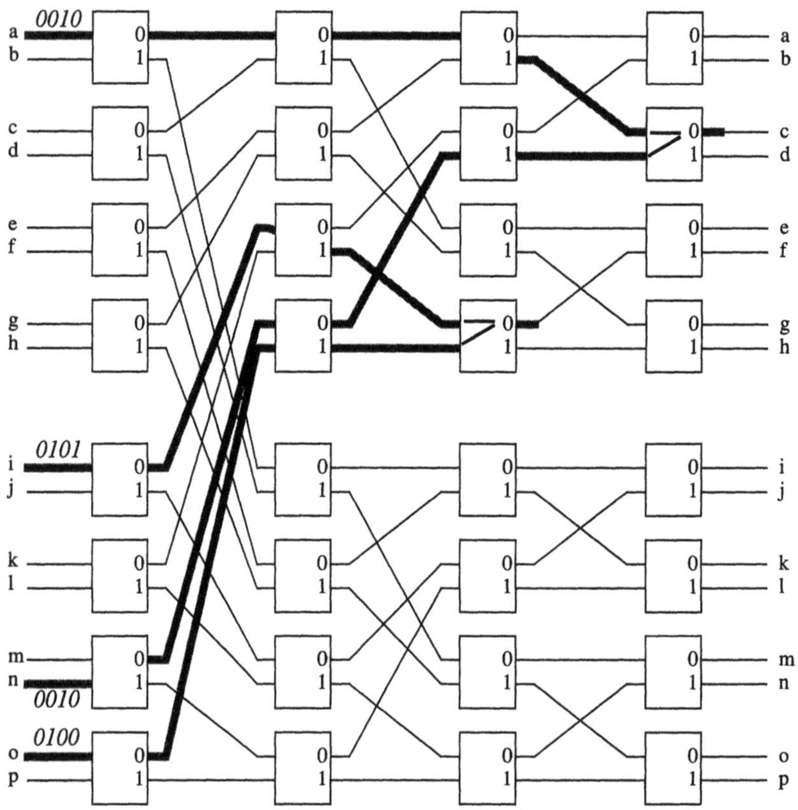

FIGURE 11.24. Four-stage delta network (collision).

Switch Controller

The switch controller that performs processes under the control of the host processor, which is located in the 4 x 4 ATM switch. The functions performed by the switch controller are described below:

- Sets the bit map conversion table in the ATM switch;
- Contains the address filter and the buffer control logics.

Shared-Memory Switch Design

A shared-memory switch is shown in Figure 11.27 and is constructed of the following functional blocks.

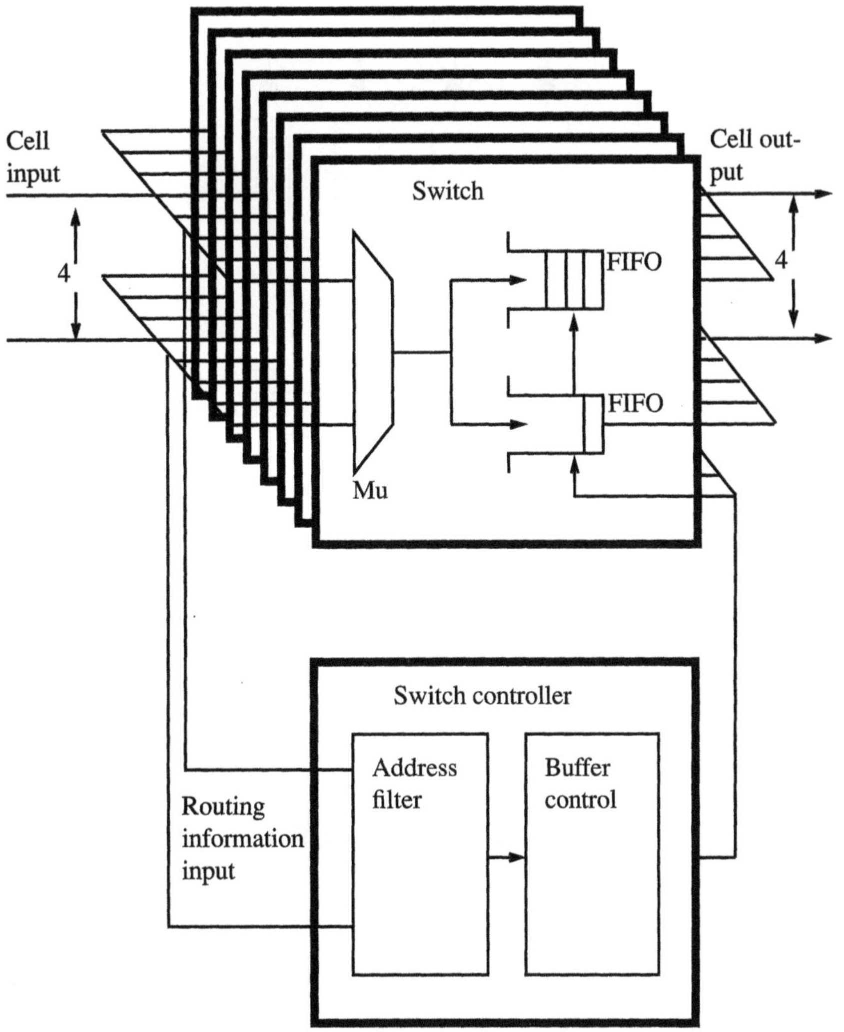

FIGURE 11.25. 4 x 4 ATM switch architecture.

Input Port

There are three input ports, numbered input port 1, input port 2, and input port 3. ATM cells T2, T1, and E3 arrive at the input port 1. The first character of each cell is the payload, and the second number is the destination output port. Therefore, E3, the third cell, arrives at this port carrying a payload of "E" directed to output port 3. Similarly, cells S3, W2, X1, and P3 arrive at the input port 2, and cells T3, E1, O2, A1, and S1 arrive at the input port 3.

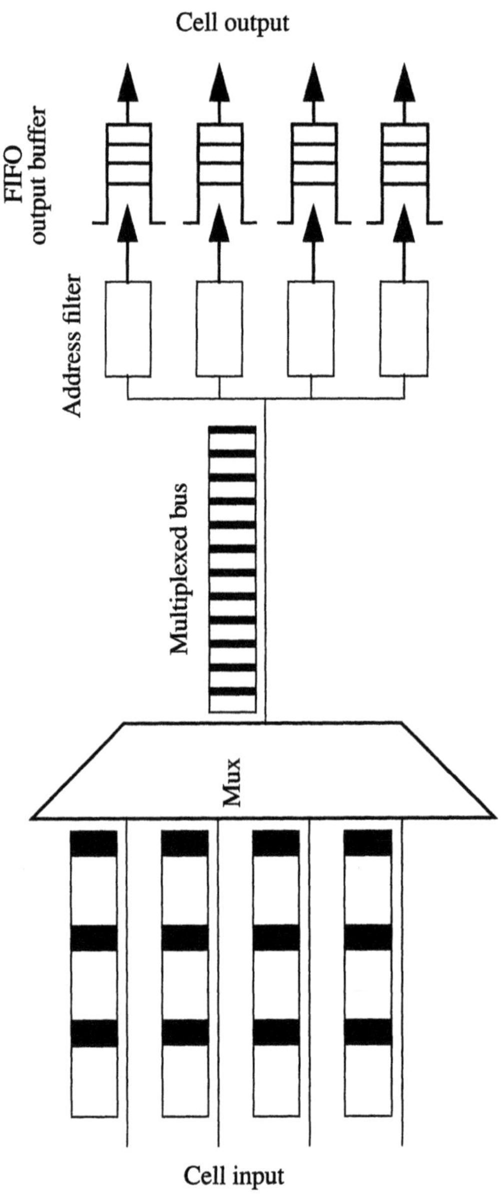

FIGURE 11.26. Switch operation.

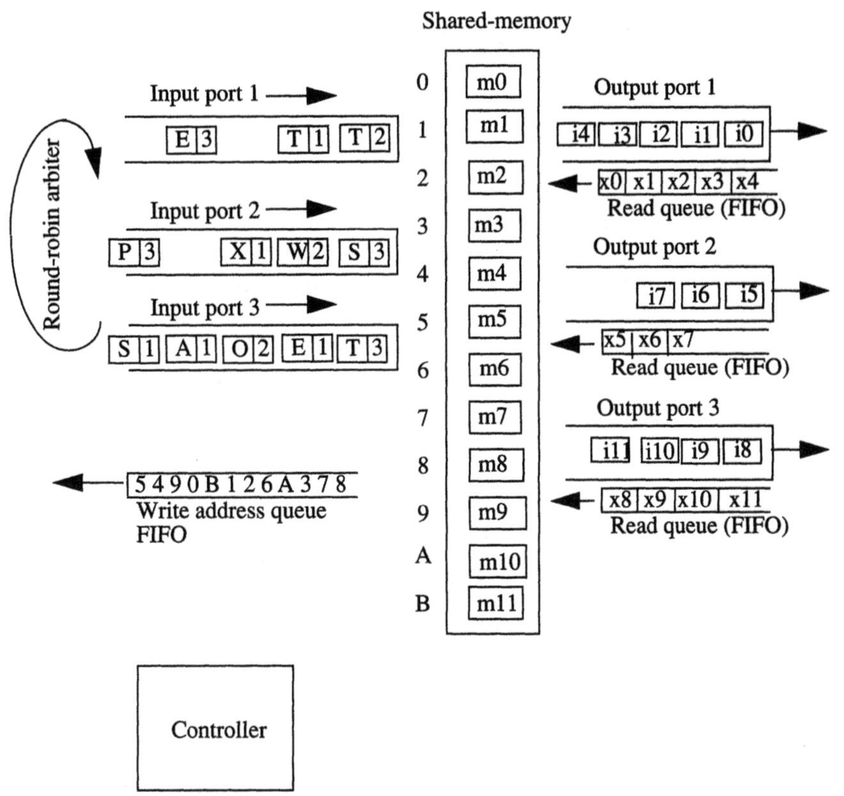

FIGURE 11.27. Example of a shared-memory switch operation.

Output Port

There are three output ports, numbered output port 1, output port 2, and output port 3. Cells received at the input ports are switched to these output ports.

Shared Memory

Shared memory stores the ATM cells temporarily before they are switched to their appropriate ports. Cells from input ports are written into the shared memory and read out to the output ports.

Write Address Queue

The write address queue is a FIFO and holds the available addresses of the shared memory where the next ATM cells can be stored. All the addresses are recycled as soon as they are freed.

Read Queue

Read queues are also FIFOs, and they hold the available addresses of the shared memory where the next ATM cells can be read out into the output ports. There are three read queues, and each of them is associated with a particular output port.

Arbitration Logic

A cyclic arbitration logic is used to determine the winning cell that is to be written into the shared memory. The cell from input port 1 is written first into the shared memory then input port 2 and then input port 3 in a round-robin fashion. If an input port has no ATM cell to be written into the shared memory then that input port's time slot is passed to the next input port.

Controller

This is the central controller for the shared memory switch. It receives instructions from the arbitration logic about which input port should be serviced next. It reads a write pointer (address) from the write address queue then it writes an ATM cell from the input port to the shared memory. It then stores this write pointer into the read queue of the appropriate output port. In the process, all the cells that arrive at the input ports are written into the shared memory and the read queues are filled with proper addresses.

When a cell is read out from the shared memory then that address is made available for the write address queue and stored in that FIFO.

The following are the contents of the shared memory, output ports, and read queues:

Contents of the shared memory (m0 to m11)

m0	m1	m2	m3	m4	m5	m6	m7	m8	m9	m10	m11
T	E	X	A	S	T	O	P	S	T	E	W

Contents of the output ports buffer (i0 to i11)

i0	i1	i2	i3	i4	i5	i6	i7	i8	i9	i10	i11
T	E	X	A	S	T	W	O	S	T	E	P

Contents of the read queue entries (x0 to x11)

x0	x1	x2	x3	x4	x5	x6	x7	x8	x9	x10	x11
0	1	2	3	8	5	B	6	4	9	A	7

CHAPTER 12 VPI/VCI Translation and Compression

Address Compression

The concept of virtual memory comes into mind when the total address space is very large and the system is unable to implement the total memory required to access all memory locations because of cost and real estate associated with such implementation. Moreover, when the addressable locations are not contiguous and the system must support any combinations of the address field (all the addressable locations), then the system is required to translate the full address field. In this case, the system will have a part of the addressable memory locations (noncontiguous) in the address processor and a relatively small memory (contiguous) implemented in the system. If the particular address of the total address space is currently in the address processor, then the address processor will generate (translate/compress) a real address to the memory, which is installed in the system. The real address space is the same as the address space of the installed memory. Figure 12.1 shows the address space versus target locations (real address).

In the case of an ATM system, the total supported address space is the concatenation of the VPI and the VCI fields of the ATM cell header. These VPI and VCI fields are used to identify ATM cells that are associated with assigned virtual connections with two levels of connection hierarchy. The VPI field of the ATM header is 8 bits wide for a UNI and 12 bits wide for an NNI. The VCI field is 16 bits wide for both the UNI and NNI. The parameter table for the particular connection is addressed by these two fields. These parameter tables contain all the information needed to process incoming ATM cells. To fully decode a VPI/VCI connection identifier using a straightforward single table lookup mechanism, one must use the 16 M table entry (conventional memory) for a UNI and 268 M for an NNI. Therefore, the cell-based ATM technology needs address processing (translation/compression) that is due to the large address space requirement. Thus, using such a mechanism to support all

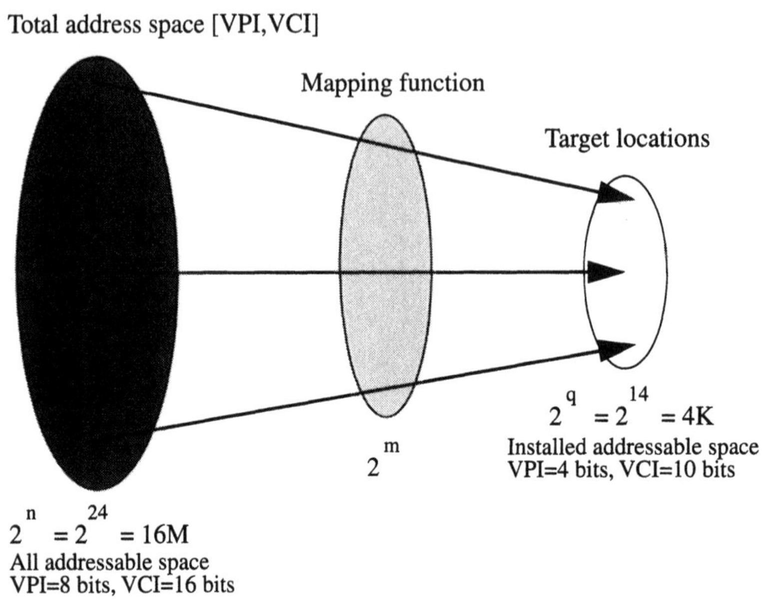

FIGURE 12.1. Address space versus target locations (real address).

possible combinations of VCI and VPI values is not practical because of the huge lookup table requirement. Although the number of total-connection identifier space is quite large, the number of actual connections at any one time is generally quite small. Figure 12.2 shows the virtual memory address concept for ATM connections.

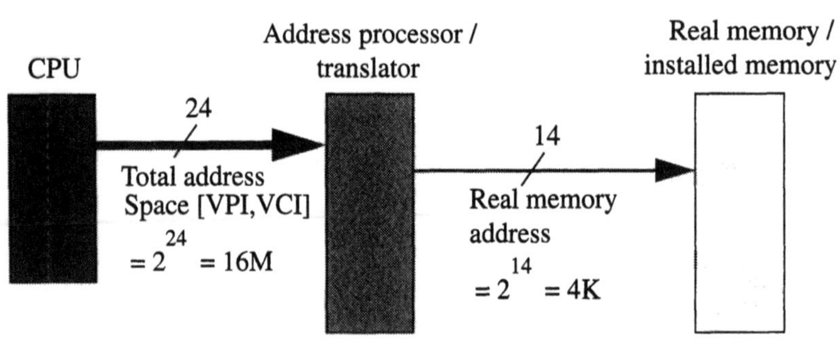

FIGURE 12.2. Virtual memory address concept for atm connections.

The ATM address processing algorithm is covered in two parts. The first part covers address translation, which is the way the incoming VPI/VCI of ATM cell header is translated into a real memory address. This points to the connection parameter table for connection information of each arriving ATM cell. The second part covers the active connection set, which is all the active connections (VPI/VCI) currently in the address processor. These will be compared with the incoming VPI/VCI of the ATM cell header to find a match and generate a real memory address. When a new connection is made, it must add that VPI/VCI into the active connection set so that it will be treated as a valid connection number (VPI/VCI). Similarly, when a connection is removed, the processor will remove the VPI/VCI from the active connection set. All these sound simple, but to do the job in real time is a difficult task.

The address translation/compression can be implemented to support two possible types of ATM connections simultaneously, which are:

1. VPC connection;
2. VCC connection.

ATM Address Translation

The address generated by the VPI/VCI field of ATM cell header is called the *virtual address*, because it may differ from the address used to access the parameter table (installed memory). The total number of connections (addressable by VPI/VCI) available in the ATM system per link is called the *virtual address space*. The address to access the parameter table (installed memory) is called the *real address*, because for each such address, a corresponding memory location really exists.

In the absence of relocation, virtual addresses are identical to real addresses. In the case of relocation, however, they may not be the same, so that the virtual addresses must be translated into the real addresses, which is called *address translation*. The translation is performed in the address processor unit.

Because of the limited size of installed memory, a VPI/VCI may not be in the installed memory (for an inactive connection). This VPI/VCI then must be added in the installed memory when a new connection is established.

The function of an address processor can be described as follows: the set of virtual addresses $VPI/VCI = \{0, 1, 5, h, m, ..., n\}$ is mapped on to the set of allocated real memory addresses $CI = \{0, 1, 2, 3, ..., q\}$ by using the function f: VPI/VCI --> CI. The f is defined as:

- $f(x) = y$ if the data at virtual address x are in CI at real address y;
- $f(x) = z$ if the data at virtual address x are missing from CI.

If $f(x) = z$, a missing VPI/VCI number occurs, the ATM cell must be extracted from the network for further processing by the onboard computer. A simple address translation mechanism is shown in Figure 12.3.

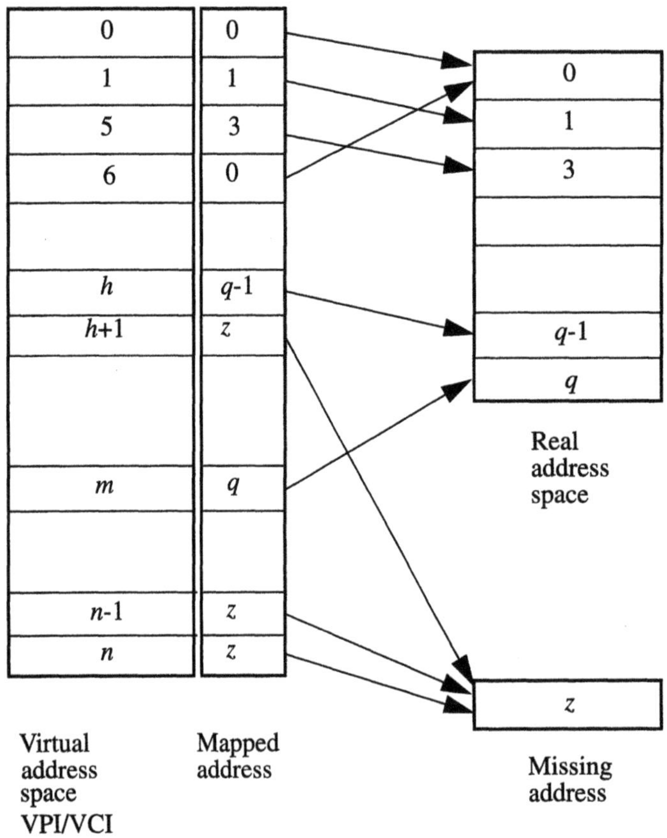

FIGURE 12.3. Simple address translation mechanism.

Active Connection Set

The active connection set is referred to the assigned VPI/VCI numbers. All the assigned VPI/VCI numbers have their own parameter tables in the installed memory; therefore, these VPI/VCI numbers can be translated into real memory addresses to point to their own parameter table. If a connection is removed, that particular VPI/VCI number will be removed from the translation list. For a permanent virtual connection, the VPI/VCI is assigned permanently. For the switched virtual connection, however, the VPI/VCI is added or deleted from the translation list in real time.

Correlation Between Maximal Number of Connection Support Versus Number of Bits Used in the VPI/VCI Field

According to standards, the system must support up to 256 virtual path connections for UNI and 4K virtual path connections for NNI. It means all 8 defined bits of the VPI field for UNI and all 12 defined bits for NNI have to be considered for address translation.

The number of VCI bits is 16 for both UNI and NNI. The maximal number of connections to be supported by an ATM system is variable. The number of bits used in the VCI field implies the number of connections supported by the system.

All the examples in this text assume the following requirements for the key (address) to be translated:

- VPI field of the ATM header: 8 bits (UNI), 0 to 256 (VPI [7:0]);
- VCI field of the ATM header: 10 bits, 0 to 1,024 (VCI [9:0]).

Software Address Translation Algorithm

Improved microprocessor architecture along with clever software algorithms have kept up with the address processing requirements. When a connection is added or deleted, a significant number of data is required to be replaced and updated. As a result, the system performance is hampered. The following high-speed software algorithms are commonly used for ATM address processing:

- Hash;
- Linked list;
- Trie tree.

Hash

The inverted page table (IPT) consists of a link, VPI, VCI, and real address (RA) fields. The link field forms a chain of IPT entries with the same hash value. Address translation is performed by hashing the VPI/VCI number of the ATM header so that it can be used to address an entry in the IPT. The content of the VPI/VCI field of that entry is then compared with the VPI/VCI of the ATM cell header. If it matches, then the RA value is taken as the selected real address. If it does not match, the link field is used to check the next IPT entry. The longer the chain, the longer it takes to translate VPI/VCI. Figure 12.4 shows the hashing process.

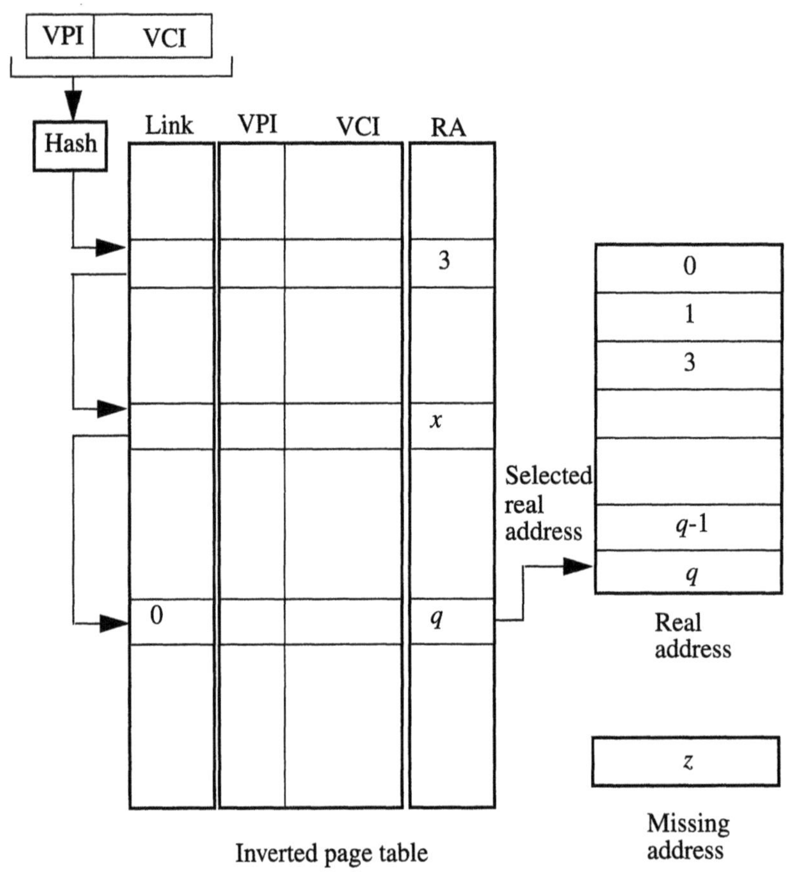

FIGURE 12.4. Addressable translation by hashing.

Linked List

In the linked list algorithm, the VPI field is mapped directly to obtain a connection number, if the connection is VPC. If the connection is VCC, then the VPI field provides the VC table address for further translation. The VCI field is split into several fields to allow for multilevel VCI translation. Figure 12.5 shows the linked list process.

Trie Tree

A basic *trie* is a combination of a simple table and a linked list. In this algorithm, the key address is segmented into a number of sections, and each section is used to address a different level of a search tree. The VPI field is mapped directly to get the

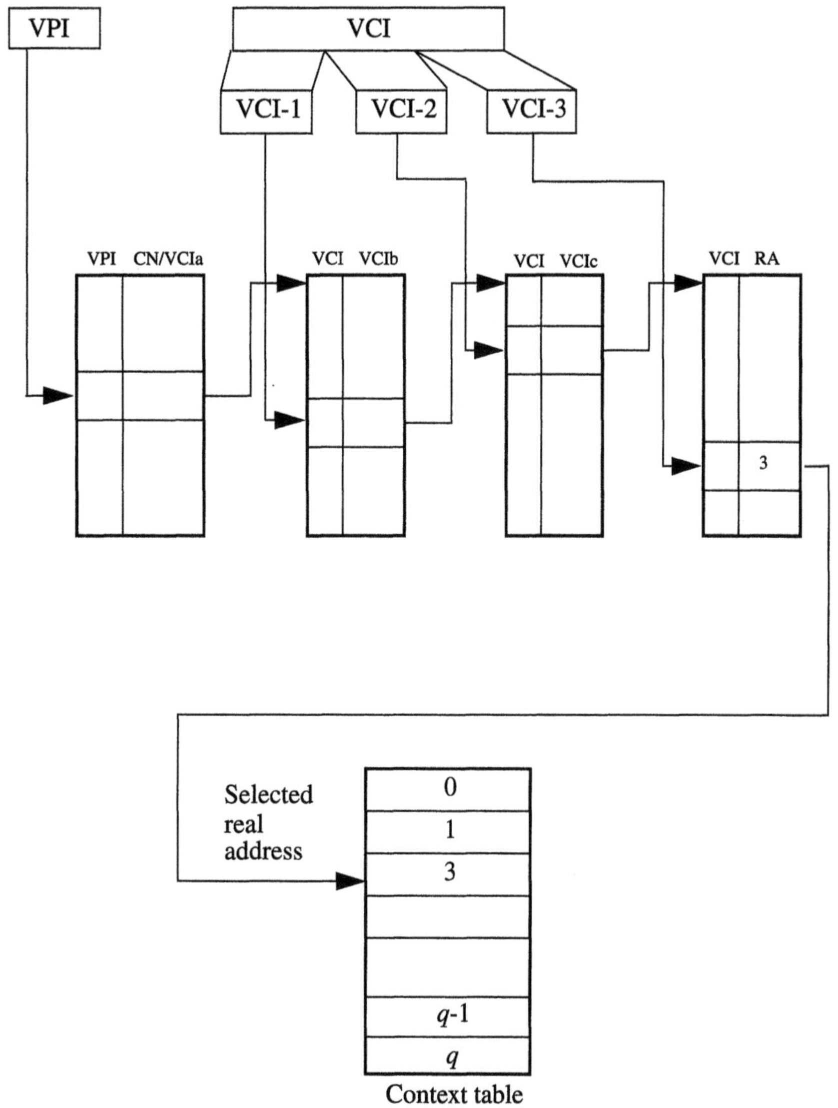

FIGURE 12.5. Addressable translation by linked list.

first VCI segment row address. Each row has *m* number of columns. The *m* is the maximal value of the VCI that each segment can hold. The first VCI segment value is the displacement from the row address that points to the appropriate column. This column contains the row address for the second VCI segment. The second VCI segment value is used to find the desired column that contains the third VCI row address. The third VCI segment value is the displacement from the row address to find the appropriate column that contains the real address. A simple searching procedure for a trie tree is shown in Figure 12.6. Figure 12.7 shows the trie tree structure for a VPI/VCI translation.

Hardware Address Translation Algorithm

The ATM cells are only 53 octets long. In the high-speed physical link, OC-3 for example, the time to pass an ATM cell is very short. Therefore, the time available to capture and process address information is relatively small. The physical layer time constraints on address processing have led to a very complex address processing alternative in the ATM network. The following equation shows the time to process an ATM cell received from an OC-3 interface:

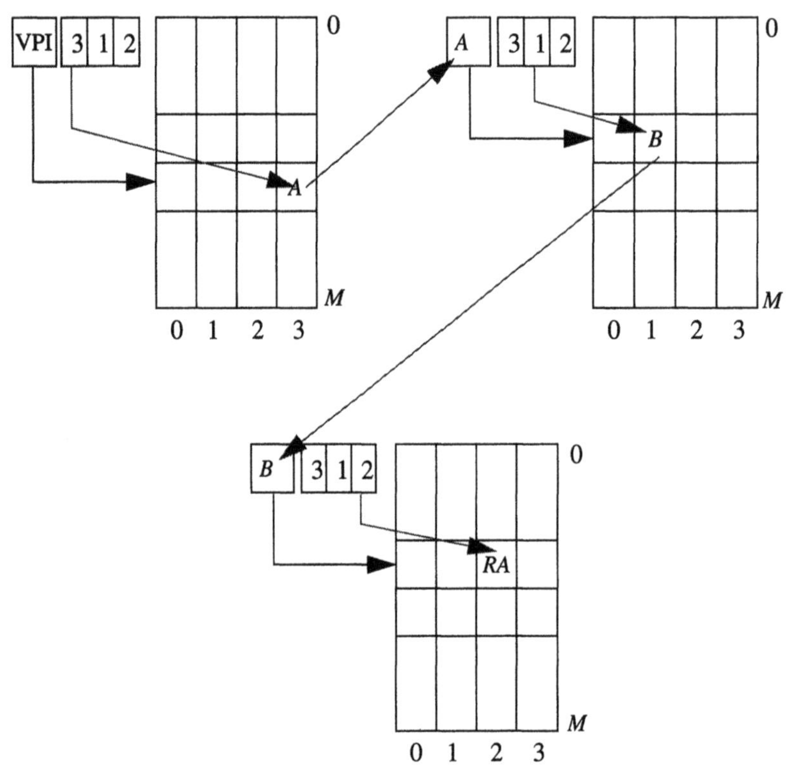

FIGURE 12.6. Simple searching procedure for a trie tree.

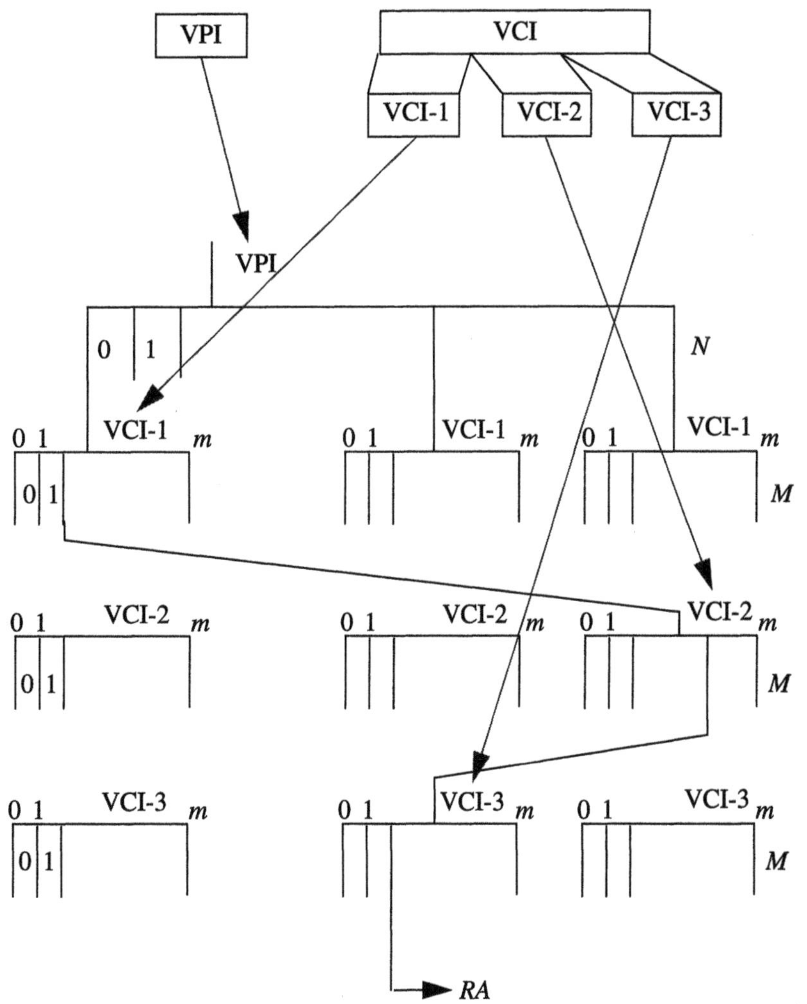

FIGURE 12.7. Trie tree structure for a VPI/VCI translation.

$$\frac{53 \text{ octets} \times 8 \text{ bits}}{155.52 \times 10^6} = 2.73 \ \mu s$$

The following equation shows the time to process an ATM cell received from an FDDI interface:

$$\frac{53 \text{ octets} \times 8 \text{ bits}}{100 \times 10^6} = 4.24 \ \mu s$$

The following equation shows the time to process an ATM cell received from a DS-3 interface:

$$\frac{53 \text{ octets} \times 8 \text{ bits}}{44.736 \times 10^6} = 9.48 \ \mu\text{s}$$

Before the evolution of broadband networks, the system operations were relatively slow, so the address translation mechanism was not critical to system operation. Software solutions were adequate for the address lookup requirements. As the broadband network started growing, the address processing throughput became the biggest challenge and software addressing schemes alone could no longer support the current requirements.

The content addressable memory (CAM) is appropriate if the utilized set (number of actual connections at any one time) is very small, because of the speed. If the cost of implementation is no object, then a CAM is appropriate. The CAM is very complex (expensive) and consumes high power in the current technology.

Content Addressable Memory

The memory used to store the VPI/VCI number and its associative real address is called a CAM, or an *associative memory*, because the desired RA is found by associating its contents with the VPI/VCI number. A parallel comparison of all entries in the CAM is made with the VPI/VCI number. When a CAM entry matches, the RA of that entry is used to access the real memory. When no match is found, the ATM cell is removed from the cell flow. Figure 12.8 shows how the CAM is used for address translation.

The real-time CAM operation can be implemented in hardware to maintain ATM link speeds. The additions and deletions of the individual entries of the CAM are implemented in software. The hardware provides the contention free arbitration between the CPU and the address compression block into the CAM space.

Figure 12.9 shows the internal circuit of a CAM. If an incoming ATM cell header contains VPI=5 and VCI=23, then the RA is 2000.

ATM Address Processing Block

ATM address processing block translates the VPI and VCI fields of the ATM cell header into a real address, which is the pointer for the connection parameter table for each connection. The VPI/VCI can be extracted from the cellstream by the ATM address processing block, or the ATM processor can provide these two fields as the key address to be translated into RA. This block must provide the RA within a cell time.

The address processing block is shown in Figure 12.10. In general, it interfaces with the following three system blocks:

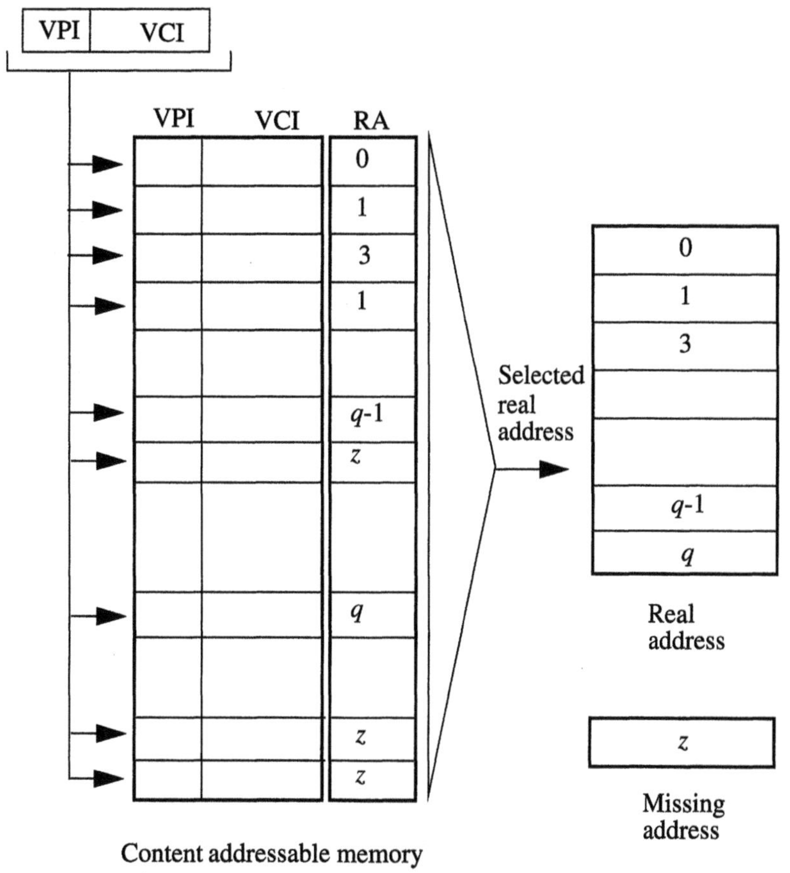

FIGURE 12.8. Content addressable memory (translation lookaside buffer).

1. CPU;
2. Address processing memory;
3. ATM processor.

CPU Interface

The CPU accesses the address processing memory through this interface to update tables (adding and deleting connections) when necessary. This interface is composed of the following signal lines:

- Address bus;
- Data bus;
- Control bus (RD, WR, CS).

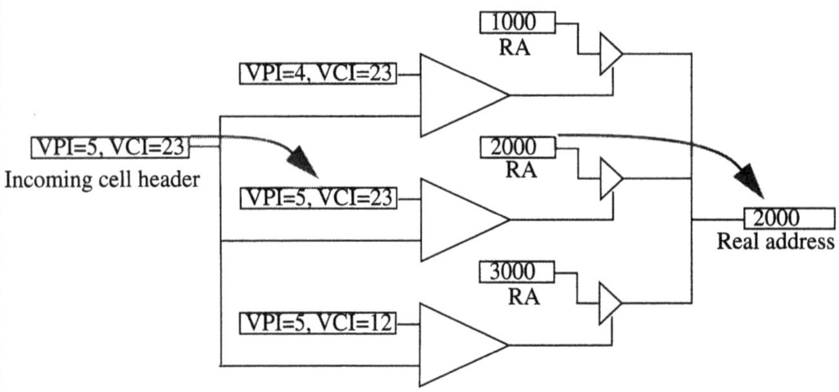

FIGURE 12.9. CAM block diagram.

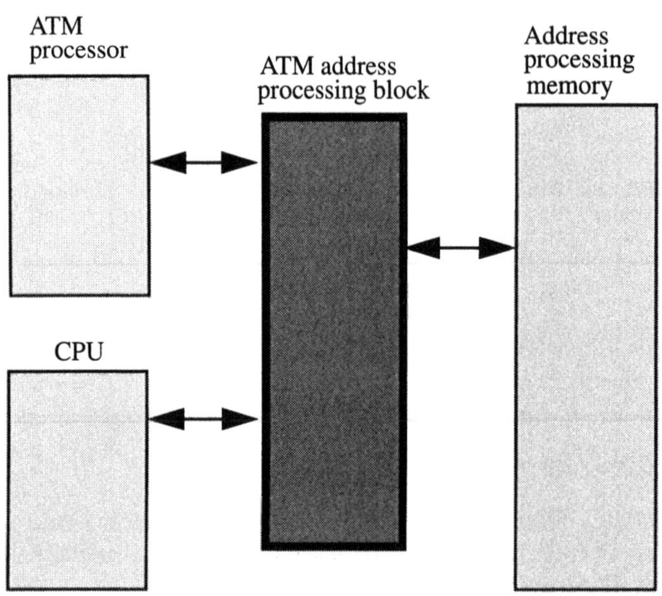

FIGURE 12.10. Address processing block.

Address Processing Memory Interface

The address compression memory contains all the entries necessary to translate VPI/VCI numbers into real addresses. It is normally a fast SRAM.

ATM Processor Interface

The ATM processor communicates with the address processing block by using the following three handshake signal lines:

1. Start;
2. Done;
3. Valid.

There are two registers located in the address processing block:

1. KEY register;
2. RA resister.

The VPI and VCI numbers are written into the KEY register by the ATM processor. The ATM address processing block starts computing the RA when the start signal is activated by the ATM processor. When the RA is computed, it is then written into the RA resister. At the same time, the done signal is also asserted. The RA is then read by the ATM processor. If the address translation is successful, then the valid signal is asserted. The valid signal means the content of the RA register is valid. The ATM processor interface registers are shown in Figure 12.11.

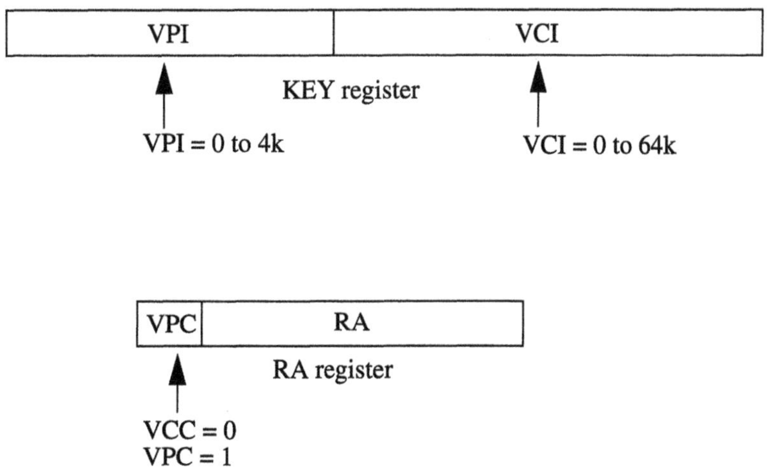

FIGURE 12.11. ATM processor interface registers.

CHAPTER 13

ATM Signaling

Signaling

We refer to the process for setting up connections across an ATM network as *signaling*. LAN technologies have been connectionless because data are transmitted without the need to set up a connection from one endpoint to another. The use of signaling is a new concept to the LAN environment. Signaling is not new, however, to the public network industry, because the public networks use very complex signaling protocols to set up and control telephone calls. Refer to ATM Forum document AF-SIG-0061.000, and ITU-T Q.2100, Q.2110, Q.2130, Q.2140, Q.2144, and Q.2931 documents for more information.

ATM standards define the following two virtual connections:

1. Permanent or provisioned virtual circuits;
2. Switched virtual circuits.

Permanent or Provisioned Virtual Circuits

Permanent or provisioned virtual circuits (PVC) are circuits that are established either manually or through some other external means (e.g., through a management protocol).

Switched Virtual Circuits

Connections set up on command by nodes are referred to as switched virtual circuits (SVC). These are supported by a signaling protocol.

The ITU-T has been working for some time on signaling and is working on a signaling protocol known as Q.2931. It is based on the N-ISDN signaling protocol Q.931. The ATM Forum has adopted some aspects of the ITU-T Q.2931 for its first phase of UNI signaling standards in AF-SIG-0061.000.

On ATM networks, addresses are used to set up connections by identifying the node or nodes with which the source node wishes to communicate. Private networks may use a variety of addresses. These addressing procedures may include 48-bit MAC addresses (used on LANs), or network layer addresses, such as IP addresses, or OSI network layer service access point (NSAP) identifiers.

This signaling is needed to support multipoint connections used for services, such as multicasting.

The ATM network signaling points are shown in Figure 13.1.

The four types of interfaces (standardized by both ITU-T and ATM Forum), are shown in Figure 13.2.

The UNI signaling is described in the ATM Forum UNI signaling specification 3.1 and the ITU-T recommendation Q.2931. The NNI signaling is described in the ATM Forum P-NNI (private-NNI) signaling specification and ITU-T recommendation B-ISDN, user part (B-ISUP).

Signaling Protocols

Signaling protocols operate across ATM interfaces using the services of the ATM network. Signaling packets are transmitted in ATM cells across reserved VPI/VCI values. A signal will be generated by a node to request that a connection be set up or torn down or to request services from the network. The request for a connection will also carry specifications pertaining to the desired class of service and the traffic

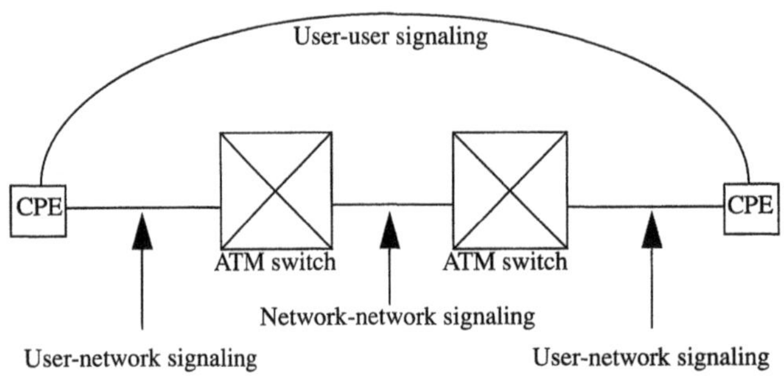

FIGURE 13.1. ATM network signaling points.

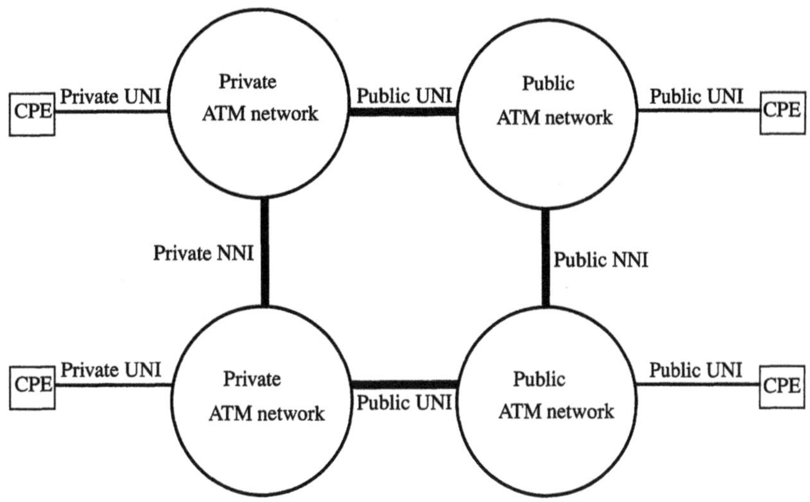

FIGURE 13.2. B-ISDN signaling interface.

management parameters. A connection request from the originating node will pass to the destination node to ask permission to set up a connection. If the destination node agrees, then the network will establish a virtual circuit across the network linking the two nodes and will map the appropriate VPI/VCI values to the intermediate ports.

The signaling standards specify exactly what services may be requested and how these requests will be coded. They do not specify, however, the manner in which a network may implement internal signaling requests. It may be simplest to pass all signaling requests across a reserved VPI/VCI value to a central signaling entity. This entity, perhaps using management protocols, would then establish circuits across the network. It may be necessary to set up connections among multiple switches. In this situation, each ATM switch or group of switches would be associated with a signaling entity. The signaling entities would then pass connection setups from one to another until a connection was made across the network from the source to the destination.

For setting up connections among multiple switches, there must be signaling standards for the entities to be able to talk to one another. An NNI signaling standard would also require an agreement on protocols that entities could use to discover the topography of the network and therefore route signaling requests to one another. The problem with network topography is analogous to the internetwork routers being used today. The problem is how to route packets across networks from one router to another. Signaling packets, as they are used to set up connections, must necessarily be connectionless. This might be one method in which current routing protocols may be applied to ATM networks.

There is still some controversy about the types of addresses to be used for ATM signaling. For instance, MAC addressing may be used to allow for MAC bridging to be performed over ATM networks. MAC addresses have no hierarchy. They are flat addressing spaces. It will be very difficult to scale MAC-address-based networks beyond a few small switches. MAC addressing could suffice during the early implementation of ATM network signaling but will have to be replaced with a hierarchical network layer address, such as E.164 or IP addresses. It is vital that an NNI signaling standard be adopted to ensure the multivendor interoperability of ATM switches and to halt the development of a multitude of proprietary signaling protocols.

The ATM/ISDN signaling protocol stack is shown in Table 13.1.

ATM UNI Signaling

ATM UNI Signaling Capabilities

ATM UNI signaling supports the following basic capabilities:

- Switched channel connections;
- Point-to-point and point-to-multipoint switched channel connections;
- Connections that require symmetric or asymmetric bandwidth;
- Single connection calls (point-to-point or point-to-multipoint);
- Basic signaling functions by any of protocol messages, information elements, and procedures;
- Class X, class A, and class C ATM transport services;
- Request and indication of signaling parameters;
- VPI/VCI assignments;
- A single, statically defined out-of-band channel for signaling messages;

TABLE 13.1. ATM/ISDN Signaling Protocol Stack

ISDN	ATM
Q.931	Q.2931
Q.921	SAAL
	ATM
I.430/I.431	SONET

- Error recovery;
- Public UNI and private UNI addressing formats for unique identification of ATM endpoints;
- A client registration mechanism for exchange of addressing information across a UNI;
- End-to-end compatibility parameter identification.

Switched Channel Connections

Switched channel connections, also referred to as demand channel connections, are established in real time using signaling procedures. These connections remain in place for an unspecified amount of time but would not automatically be reestablished after a network failure. Permanent connections are those that are set up and torn down by way of provisioning. These connections generally are in place for long periods of time and would be automatically reestablished after a network failure.

Point-to-Point and Point-to-Multipoint Connections

A point-to-point connection is a set of associated ATM virtual channel or virtual path links that connect two endpoints. A point-to-multipoint connection is a set of associated ATM VC or VP links with the properties listed below.

One ATM link serves as the root link in a simple tree topology. Each of the other nodes in the connection is a leaf node, and all receive copies of the information sent by the root link. The leaf nodes do not communicate with one another directly. A distributed implementation can be used to connect leaves to the tree.

A point-to-multipoint connection is made by first establishing a point-to-point connection between the root node and a leaf node. Other leaves are added on the basis of an "add party" request from the root node. A leaf node may be added or dropped from this connection at any time by a request from either the leaf node or the root node.

Connections With Symmetric or Asymmetric Bandwidth

Point-to-point bidirectional connections may have the bandwidth specified in both directions, forward from the calling party and backward from the party being called to the calling party.

Single Connection per Call

Calls with one and only one connection are supported, but the connection may be point-to-point or point-to-multipoint. The basic signaling functions are supported by the following protocols:

- Connection/call setup;
- Connection/call request;
- Connection/call answer;

- Connection/call clearing;
- Reason for clearing;
- Out-of-band signaling.

Class X, Class A, and Class C ATM Transport Services

Class X service is a connection-oriented ATM transport service in which the AAL, traffic type (VBR or CBR), and timing requirements are defined by the user. The user selects the bandwidth and QoS in his or her setup message to establish the connection.

Class A service is a connection-oriented, CBR service. It has end-to-end timing requirements. As a result, it may require strict cell loss, cell delay, and cell delay variation performance. As with class X, the user determines the bandwidth and QoS in his or her setup message.

Class C service is a connection-oriented VBR ATM transport service. There is no end-to-end timing requirement, and once again the user determines the desired bandwidth and QoS in his or her setup message.

Class D service is not directly supported by phase 1 signaling. It can be supported by way of class X or class C connection to a connectionless server.

Signaling Parameter "Request and Identification"

The user sends a setup message indicating a value for each parameter. The receiver indicates whether the specified values can be accommodated.

VPI/VCI Support

The VPI/VCI is the way of identifying the virtual path across the UNI.

Single Signaling Virtual Channel

The point-to-point signaling virtual channel (VCI=5, VPI=0) is used for all signaling.

Error Recovery

Signaling supports error recovery.

Public UNI and Private UNI ATM Addressing

Signaling supports a number of ATM address formats that are used across the public and private UNIs to identify the endpoints of an ATM connection.

Client Registration

Signaling supports a mechanism for the exchange of address and identifier information between an end system and a switch across a UNI.

End-to-End Compatibility Parameter Identification

The following end-to-end parameters can be specified on the basis of each connection:

- The AAL type;
- The method of protocol multiplexing;
- For VC-based multiplexing, a protocol that is built in and taken from a list of known routed protocols or bridged protocols.

ATM Addressing

An ATM private and public network address uniquely identifies an ATM endpoint for switched virtual connection. The ATM address format is modeled after the OSI (NSAP model, as specified in International Organization for Standardization (ISO 8348) and ITU-T X.213 and E.164 standards. The ATM address format is shown in Figure 13.3.

Authority and Format Identifier

The authority and format identifier (AFI) codes are shown in Table 13.2. The data country code (DCC) specifies the country in which an address is registered. The international code designator (ICD) identifies an international organization. E.164 specifies integrated services digital network (ISDN) numbers.

Domain Specific Part Format Identifier

The DFI field specifies the structure, semantics, and administrative requirements for the remainder of the address.

Administrative Authority

The AA field specifies the organizational entity for allocation of addresses in the remainder of the domain-specific part (DSP), such as an ATM service provider, a private ATM network, or an ATM vendor.

TABLE 13.2. AFI Codes

AFI	Format
39	DCC ATM format
47	ICD ATM format
45	E.164 ATM format

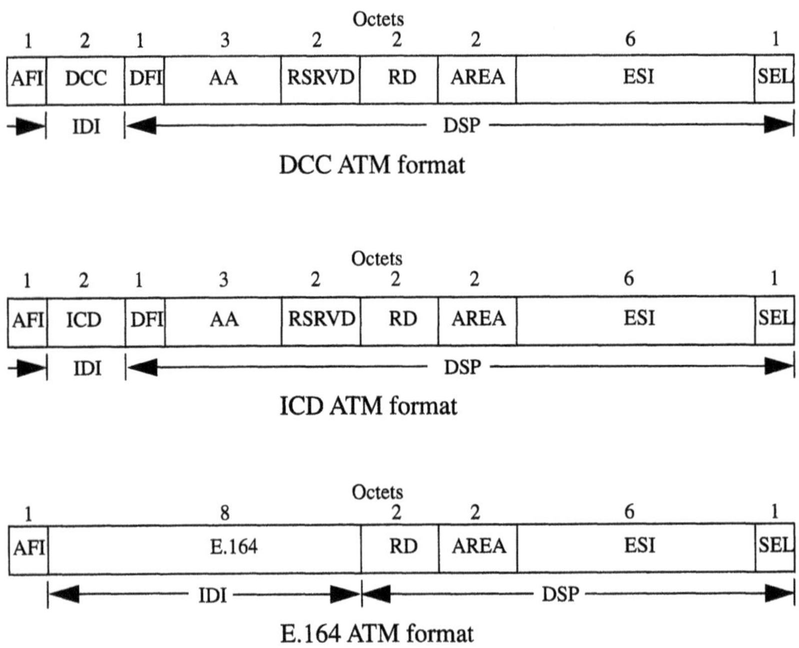

AA: administrative authority
DCC: data country code
DFI: DSP format identifier
AFI: authority and format identifier
RSRVD: reserved
RD: routing domain
ESI: end system identifier
SEL: selector

IDI: initial domain identifier
DSP: domain-specific part
ICD: International code designator

FIGURE 13.3. ATM address formats.

Routing Domain

The RD field identifies a unique domain within one of the following: E.164, DCC/DFI/AA, or ICD/DFI/AA.

Area

The AREA field identifies a unique area within a routing domain.

End System Identifier

The ESI field is globally unique, such as an IEEE MAC address (48 bit).

Selector

The SEL field is not used in the ATM routing.

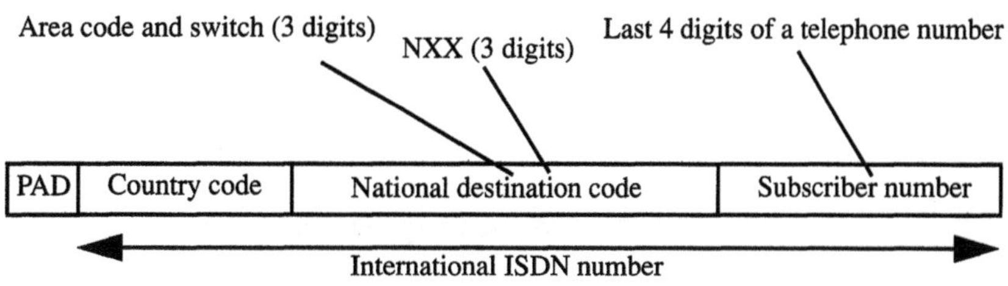

FIGURE 13.4. ITU-T E.164 numbering plan format.

An example of the E.164 (e.g., telephone numbers) numbering plan is shown in Figure 13.4. The E.164 number is a 15 digit number coded in Binary Coded Decimal (BCD). Pad digit is always a zero. Country code ranges from 1 to 3 digits. The destination code and subscriber number are translated within a country.

ATM Signaling Message Structure

Standards define signaling message structure, which highlights the functional definitions and information content of each message. Refer to ITU-T Q.2931 and ATM Forum AF-SIG-0061.000 for more information.

Messages for ATM Point-to-Point Call and Connection Control

Messages for ATM point-to-point call and connection control are listed below:

Call establishment messages:

- Call proceeding;
- Connect;
- Connect acknowledge;
- Setup.

Call clearing messages:

- Release;
- Release complete.

Miscellaneous messages:

- Status;
- Status inquiry.

The point-to-point connections shown are in Figure 13.5.

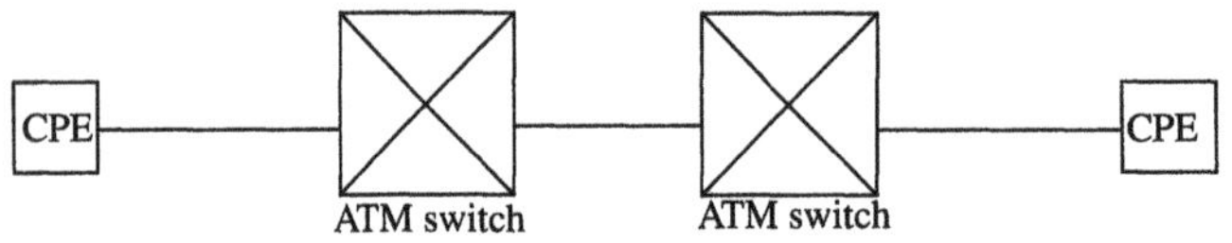

FIGURE 13.5. ATM point-to-point connection.

Messages Used With the Global Call Reference

Messages used with the global call reference are listed below:

- Restart;
- Restart acknowledge;
- Status.

Messages for ATM Point-to-Multipoint Call and Connection Control

Messages for ATM point-to-multipoint call and connection control are listed below:

- Add party;
- Add party acknowledge;
- Add party reject;
- Drop party;
- Drop party acknowledge.

The point-to-multipoint connections are shown in Figure 13.6.

A point-to-point signaling message flow is shown in Figure 13.7.

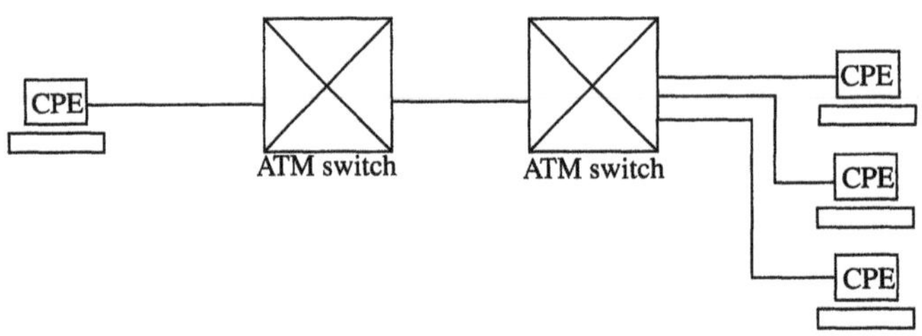

FIGURE 13.6. ATM point-to-multipoint connection.

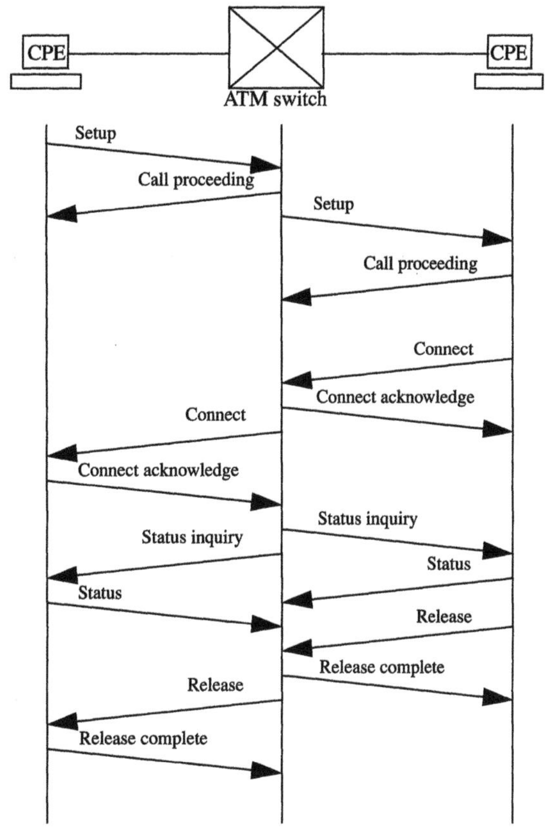

FIGURE 13.7. Point-to-point signaling message flow.

ATM Call Control States

Standards define the basic call control states for ATM calls, such as follows:

- Null;
- Call indication;
- Outgoing call proceeding;
- Call delivered;
- Call present;
- Call received;
- Connect request;

- Incoming call proceeding;
- Active;
- Release request;
- Release indication;
- Call abort;
- Restart request;
- Restart.

The call originating and call receiving states are shown in Figure 13.8 and Figure 13.9, respectively.

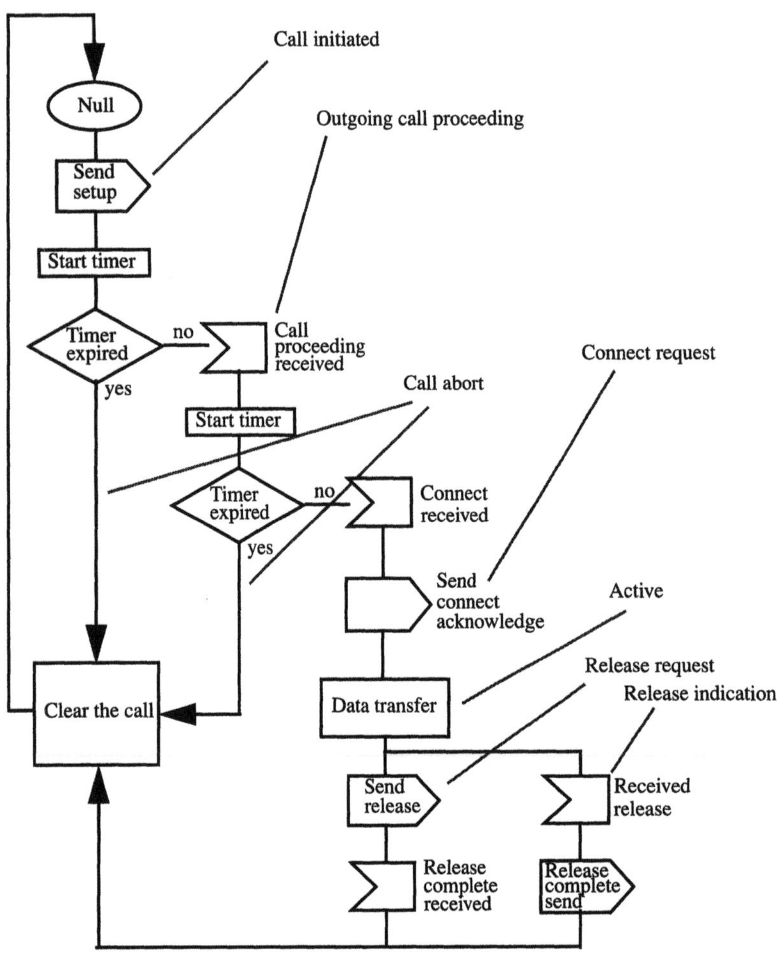

FIGURE 13.8. Call originating states.

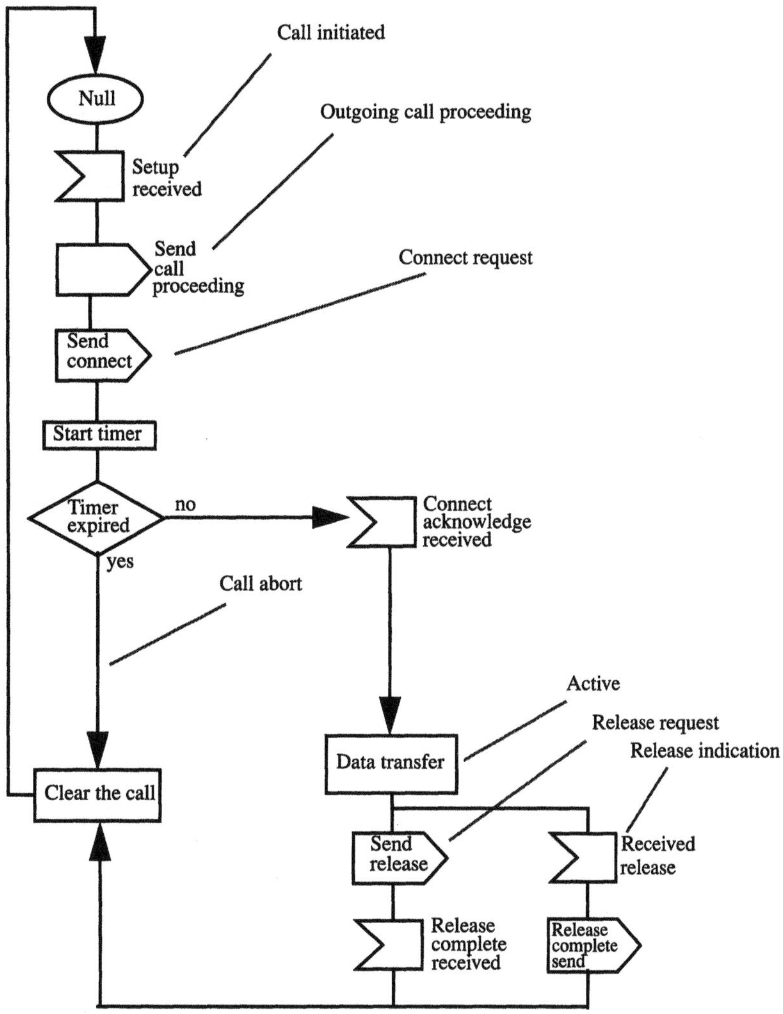

FIGURE 13.9. Call receiving states.

Multicasting

We define multicasting as the ability of one node to send a single packet to a number of other nodes. At present, this function is supported on the UNIs. These are based on broadcast media, which means every node sees every packet. A result of this function is the development of a number of LAN protocols that use broadcasting or multicasting services. An example would be address resolution protocols, such as ARP. Multicasting is used to support services like videoconferencing. The multicasting function is not basic to the ATM network, because the ATM network is both connection-oriented and based on point-to-point links. The receiving nodes only see the packets they have agreed to receive.

To support multicasting on an ATM network, there needs to be two mechanisms established. There also needs to be a method for establishing some sort of multipoint connection among all nodes in a multicasting group and a method for a single packet to be transmitted from one source and delivered to every node. This requires that somewhere in the network that packet or the cell carrying it must be replicated. Currently, some limited ability to support multipoint connection is available with UNI signaling.

The issue of replication is one for switch architects. Replication is not possible on most matrix type ATM fabrics. Cells cannot be replicated and routed at the same time. This would seem to call for a "replicating fabric" prepended to the switch matrix. Back plane-based switches, which are bus-based, can perform multicasting more simply because as a bus, each node sees all traffic and can simply copy multicast cells.

A more fundamental problem with multicasting is that AAL-5 does not support a multiplexing capability. This makes it impossible to establish a single VPI/VCI value for a multicast group (on which all nodes can both send and receive), because if a node were to receive an SDU packet from two or more members of the group simultaneously, it would not be able to differentiate the cells from the different SDUs. One means of overcoming this problem might be to identify a multicast group with a particular VPI value and use separate VCI values to identify each transmitting node in the group.

An alternative might be to use an external multicast server. This server would receive packets from all nodes in the group. The multicast server would then forward the packets to all nodes in the group, ensuring that all packets were serialized so there would be no intermingling of cells. The appropriate addressing and signaling mechanisms will be required to allow nodes to identify a multicast group and members within that group. The network will require signaling mechanisms to program ATM switches to support multicast operations.

ATM LAN Emulation

ATM connectionless service framework supports legacy LANs and SMDS networks across the ATM networks using its multicasting capability. All LAN clients and the server communicate over the ATM virtual connections (VCC). Normally, an emulated LAN is either ethernet (IEEE 802.3) or token ring (IEEE 802.5). The basic configuration of ATM LAN emulation (LANE) service is shown in Figure 13.10.

Private NNI Signaling

Private networks use their own proprietary signaling to establish or remove connections between their switches. The interoperability between two private networks requires the standardization of the private node-to-node interface. The ATM Forum

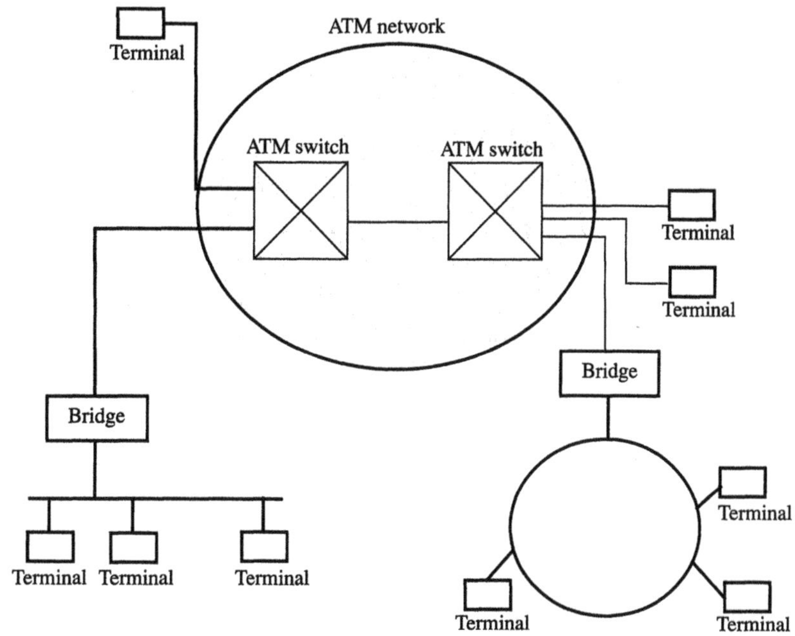

FIGURE 13.10. ATM LAN emulation.

P-NNI signaling implementation agreement specifies the procedures and messages used to dynamically establish, maintain, and remove ATM connections at the P-NNI between two private networks. The signaling protocol stack for P-NNI is shown in Figure 13.11.

P-NNI signaling is built on UNI signaling (e.g., a large portion of the P-NNI signaling messages are the same as in UNI signaling messages). P-NNI signaling procedures are symmetric, whereas UNI signaling procedures are asymmetric.

Public NNI Signaling

Public Network

The service provider networks, known as public networks, are generally constructed hierarchically and are quite different than the private networks. Users are connected to the class 5 switch (e.g., end office). These switches are connected to the long-distance carriers (e.g., AT&T, MCI, Sprint) for complete end-to-end interconnection. The long-distance networks are set up in a hierarchical fashion and are

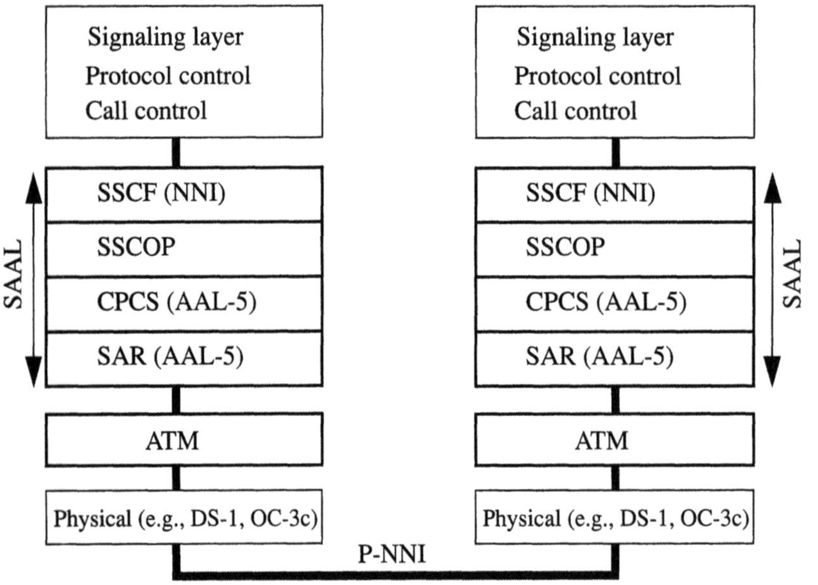

SSCF: signaling-specific coordination function
SSCOP: service-specific connection-oriented protocol
SAR: segmentation and reassembly
CPCS: common part convergence sublayer
SAAL: signaling AAL

FIGURE 13.11. Signaling protocol stack for P-NNI.

composed of toll switches, class 2 or 3 switches (e.g., primary and sectional switches), and a regional switch.

The NNI signaling standards are vital to allow a hop-to-hop connection setup, which is required by a multicast operation. It is unlikely that the connection setup by way of centralized management systems will carry ATM development very far. The full development of NNI signaling is critical to the development of true multi-vendor interoperability of ATM switches. Switches that are in the higher level in the hierarchy carry more traffic. The structure of the public network in general is shown in Figure 13.12.

Signaling messages are used to establish, release, and manage end-to-end connections. Traditionally, in-band signaling is used to set up connections (e.g., the signaling information is carried along with the user data). In this way, the signaling information to set up a connection can be transmitted only after a channel is set to the next switch along the connection path. This process accumulates delay to set up end-to-end connections.

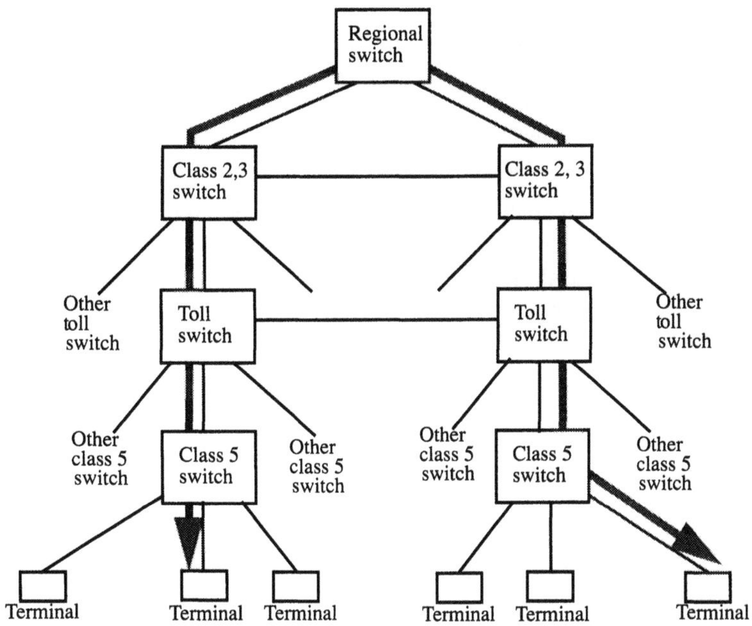

Shaded arrow shows the worst case signaling path between two terminals.

FIGURE 13.12. Public network structure.

Common Channel Signaling

Common channel signaling (CCS) was introduced to avoid the problems stated above. CCS is out-of-band signaling, meaning it uses separate signaling networks (e.g., other than voice networks) to establish end-to-end connections. The CCS networks are shared by a large number of channels.

Processors within telephone switches transfer signaling information between them through a specialized data communication network (e.g., CCS network). The standard for CCS network is defined by ITU-T recommendations on Signaling System number 7 (SS7).

The CCS network is composed of some of the following elements:

- Service switching point (SSP);
- Service control point (SCP);
- Signaling transfer point (STP);
- Switched management system (SMS);
- Intelligent peripheral (IP).

The basic CCS network is shown in Figure 13.13.

The SSP is a network switch containing table-based information and other software-based functions. The caller provides information to be looked up in the SSP tables for a match. A specific action is triggered in the SSP when a match is found with a table entry. These actions send and receive query messages to the SCP containing the service application (e.g., 800 number translation).

The SCP hosts different application software that is processed in real time. It provides reliable access to and processing of database inquiries. The 800, 900, and credit card calls are translated here depending on where the call originated. The routing information of the call will be sent back to the requesting SSP.

The STP is a packet switch that routes signaling messages among the various CCS network elements. Each message contains the destination address.

The SMS provides a framework by which new services can be added to, deleted from, or modified in the CCS network.

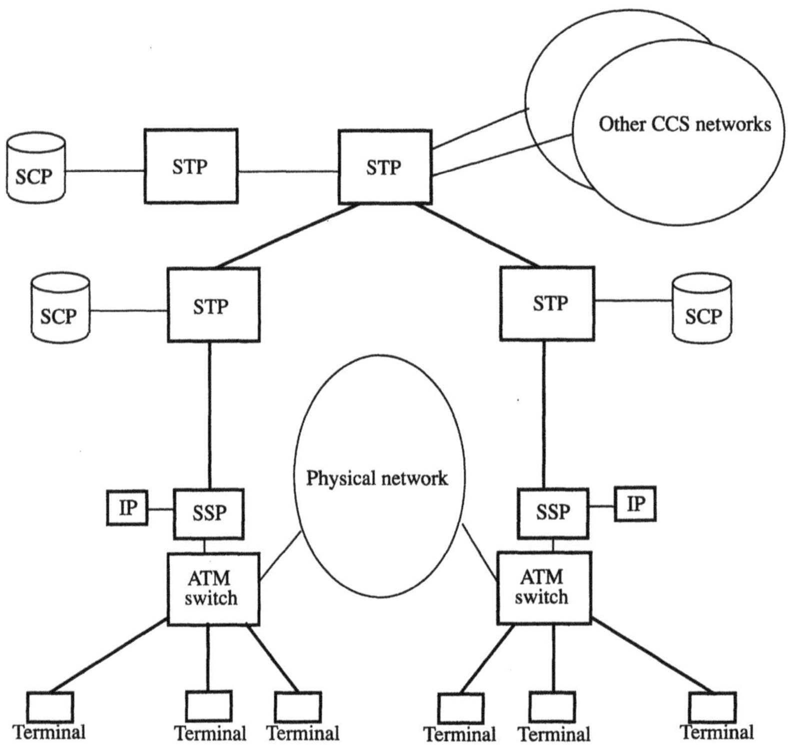

FIGURE 13.13. Basic CCS network.

The IP is a stand-alone computer that provides functions as a result of a command from the SSP. Some of these functions are: announcement messages, voice recognition, and speech processing.

SS7 Architecture

The SS7 protocol stack is shown in Figure 13.14. It is composed of the message transfer part (MTP) and the user part. The user part is composed of functions of a particular type of a user that is part of the CCS network. The MTP supports reliable transfer of the signaling messages.

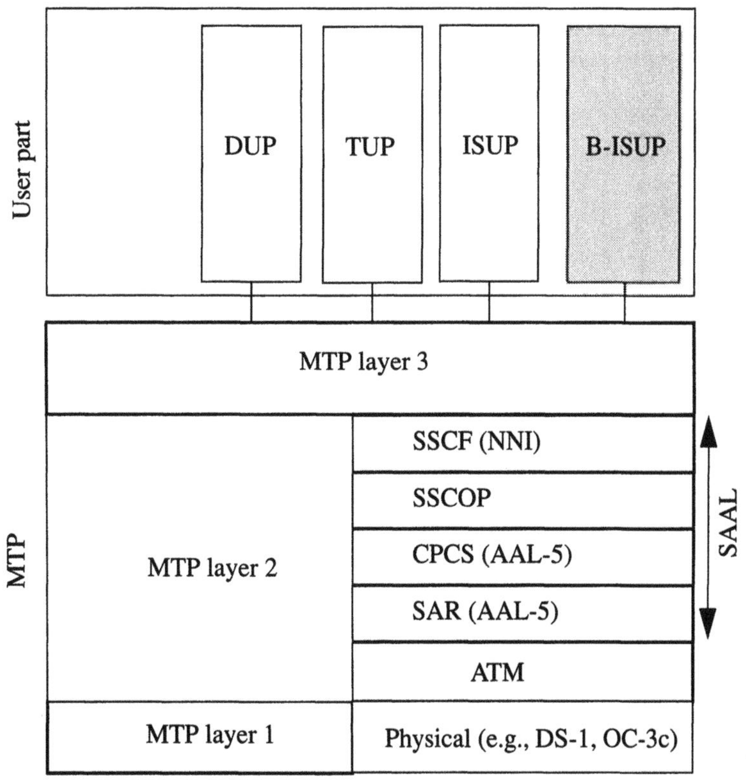

SSCF: signaling specific coordination function
SSCOP: service specific connection-oriented protocol
SAR: segmentation and reassembly
CPCS: common part convergence sublayer
SAAL: signaling AAL

FIGURE 13.14. SS7 protocol stack for public NNI.

The data user part (DUP) controls the data call facility registration and cancellation for interexchange circuits. The telephone user part (TUP) controls telephone signaling functions for international calls (e.g., signaling messages, encoding of these messages, switch performance). The ISDN user part (ISUP) controls ISDN network signaling functions for ISDN calls (e.g., signaling messages, encoding of these messages, switch performance).

The broadband ISDN user part (B-ISUP) is developed for international application as an NNI (e.g., B-ISDN applications). It is also suitable for national public network-to-node interface.

The functional description of B-ISUP is described in ITU-T recommendation Q.2761. The B-ISUP signaling information elements and their functions are defined in ITU-T recommendation Q.2762. Formats and codes of B-ISUP messages are specified in ITU-T recommendation Q.2763. The SS7 B-ISUP basic call procedures and supplementary services are described in ITU-T recommendation Q.2764 and ITU-T recommendation Q.2730, respectively.

CHAPTER 14

ATM Traffic

Traffic Objectives

The ATM network is designed to transport a wide variety of traffic classes. Any network is subject to congestion, hence traffic control is necessary. Congestion is a state in which the network may not be able to meet network performance objectives for the already-established connections. Congestion can be caused by:

- Unpredictable surge of traffic flows;
- Network fault.

The objective of traffic control is to avoid network congestion and control its effect, if it occurs. Congestion can be controlled to minimize the intensity, spread, and duration of congestion.

ATM network operates with high link speed ranging from hundreds of megabits per second to potentially gigabits per second. The requirements to support multiple services and high-speed links make traffic management in an ATM network quite difficult.

Despite the experience gained from the circuit switch network and from the packet switch network for a long period of time, traffic control in the ATM network is still relatively new and has many unresolved issues.

Source Traffic Characteristics

ATM networks are expected to support a diverse set of applications. To design and develop effective ATM traffic control methods, it is essential to understand the source traffic characteristics and its QoS requirements. Figure 14.1 shows the characteristics of different types of traffic.

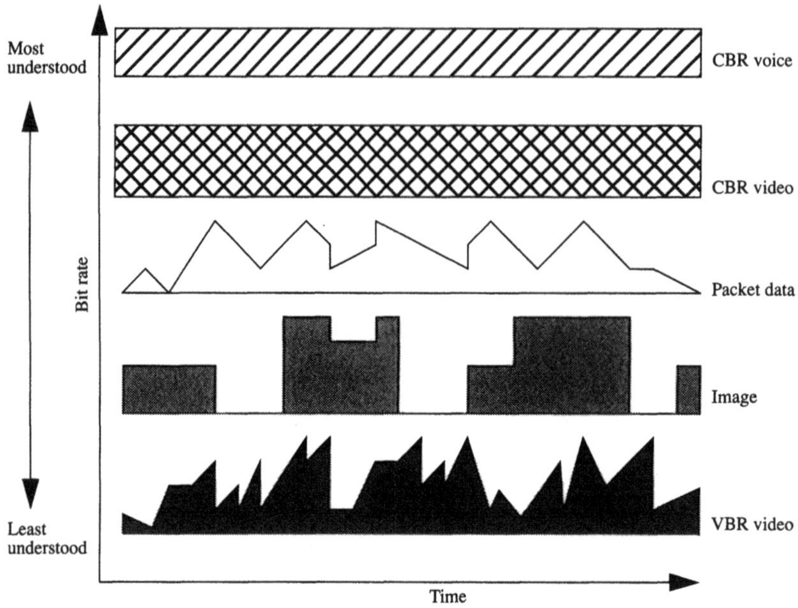

FIGURE 14.1. Source traffic of the ATM network. (*After:* Onvural, O. Raif, *Asynchronous Transfer Mode Networks—Performance Issues*, Artech House, 1994.)

The sources of the voice traffic characteristics have been studied for several decades in the context of telephone networks and are well understood. The CBR video sources produce bitstream at a constant rate. The quality of CBR video services increases at the same rate the data submitted to the network increases. The sources producing data packets are somewhat understood but are unpredictable. The image and VBR video communications are relatively new, and knowledge of their source behavior is limited and based on few system implementations.

ATM Service Architecture

According to ATM Forum (e.g., AF-TM-0056.000), ATM layer service architecture consists of the following five service categories:

1. Constant-bit rate (CBR);
2. Real-time variable-bit rate (RT-VBR);
3. Non-real-time variable-bit rate (NRT-VBR);

4. Unspecified bit rate (UBR);
5. Available bit rate (ABR).

These service categories are discussed later in this chapter.

ITU-T recommendation I.211 and ATM Forum specification AF-TM-0056.000 identify some applications of the above service categories.

CBR traffic:

- Interactive audio (e.g., telephone);
- Audio distribution (e.g., radio);
- Audio retrieval (e.g., audio library);
- Interactive video (e.g., videoconference);
- Video distribution (e.g., television, distance learning);
- Video retrieval (e.g., video on demand).

RT-VBR traffic:

- Statistical multiplexing of CBR traffic listed above and transmitting them at a variable rate to efficiently use network resources.

NRT-VBR traffic:

- Frame relay interworking;
- Response-time critical transaction (e.g., airline reservations, banking, stock market).

UBR traffic:

- Interactive text, data, and image transfer (e.g., credit card verification, banking transaction);
- Text, data, and image messaging (e.g., e-mail, fax);
- Text, data, and image distribution (e.g., news, weather report);
- Text, data, and image retrieval (e.g., file transfer);
- LAN (e.g., LAN emulation);
- Remote terminal (e.g., home office).

ABR traffic:

- Any UBR traffic listed above using ABR flow-control protocol;
- Data communications.

Quality of Service

ATM connections have a connection attribute called QoS that determines how end-to-end traffic is handled by the ATM network elements. Refer to ITU-T recommendation I.356 for more information.

QoS Parameters

In general, QoS is defined in terms of the measurements listed below for an end-to-end connection. The following QoS parameters are negotiated during connection setup between the end system and the network:

- Cell delay variation (CDV);
- Cell delay variation tolerance (CDVT);
- Cell transfer delay (CTD);
- Cell loss ratio (CLR).

The following QoS parameters are not negotiated during connection setup between the end system and the network:

- Cell error ratio (CER);
- Severely errored cell block ratio (SECBR);
- Cell misinsertion rate (CMR).

Cell Delay Variation

ATM cells of a given connection may be delayed when cells from two or more ATM connections are multiplexed. Cells are delayed because physical layer overhead are added, OAM cells are inserted, or cells of another ATM connection are inserted. CVD is a traffic parameter that is measured as cell clumping. CVD can be computed at a single point against the nominal intercell spacing, or from a traffic entry to exit point.

Cell Delay Variation Tolerance

CVDT is a traffic parameter that specifies the CDV allowed for a conforming ATM connection. CDVT is defined in relation to the PCR. Delays experienced by succeeding cells vary, however, resulting in uneven cell spacing. The upper bound on the clumping measure is called CDVT.

Cell Transfer Delay

CTD is caused by processing and queuing delay at the ATM CPEs, the network, and the switches. Processing delay is caused by coding and decoding delay, segmentation and reassembling delay, and switching and routing delay in the ATM switches. Queuing delay is caused by network congestion. In addition to the queuing and processing delays, propagation delays in long transmission links can also be a significant factor in CTD.

Cell Loss Ratio

CLR is described as the ratio of the number of lost cells that do not reach the destination to the total number of transmitted cells from the originating user.

$$\text{Cell loss ratio} = \frac{\text{Lost cells}}{\text{Transmitted cells}}$$

Cell Error Ratio

CER is described as the ratio of the number of errored cells that arrive at the destination with error in the payload to the total number of successfully transmitted cells from the originating user.

$$\text{Cell error ratio} = \frac{\text{Errored cells}}{\text{Successfully transmitted cells} + \text{Errored cells}}$$

Severely Errored Cell Block Ratio

A cell block is defined as a sequence of *N* cells transmitted consecutively on a given connection. Therefore, SECBR is described as the ratio of more than *M* cell blocks with errors (e.g., errored, lost, or misinserted cells) detected in a received *N* cell block.

$$\text{Severely errored cell blocks} = \frac{\text{Severely errored cell blocks}}{\text{Total transmitted cell blocks}}$$

Cell Misinsertion Rate

CMR is described as the ratio of the number of the misinserted cells that arrive at the destination but were not sent by the originator (e.g., wrong address) to a specified time interval.

$$\text{Cell misinsertion ratio} = \frac{\text{Misinserted cells}}{\text{Time interval}}$$

QoS Classes

The ATM Forum UNI 3.1 specification defined the following two types of QoS classes:

1. Specified QoS classes;
2. Unspecified QoS classes.

Specified QoS Classes

For connections with specified QoS class, a QoS class is specified as a part of the traffic contract.

The following four basic classes of traffic (class A-D) are included in the specified QoS class:

- Class A: CBR, circuit emulation, connection-oriented, synchronous traffic (e.g., uncompressed voice or video);
- Class B: VBR, connection-oriented, synchronous traffic (e.g., compressed voice and video);

- Class C: VBR, connection-oriented, asynchronous traffic (e.g., X.25, Frame Relay);
- Class D: VBR, connectionless packet data (e.g., LAN traffic, SMDS).

Unspecified QoS Classes

In the unspecified QoS class, the traffic parameters are not specified for VPC and VCC. The user does not expect a performance commitment from the network. An example of this class is the best effort or UBR service, where the user does not specify any traffic parameters for the intended connection.

Traffic Contract

The negotiated characteristics of an ATM connection are specified by the traffic contract. A separate traffic contract is needed for every VPC or VCC. The traffic contract is an agreement between a user and a network (e.g., service provider), that specifies the QoS, traffic descriptors, and conformance definition.

Traffic Parameters

The traffic characteristics of an ATM connection are described by traffic parameters. Some of the traffic parameters are:

- Peak cell rate (PCR);
- Sustainable cell rate (SCR);
- Maximum burst size (MBS);
- Burst tolerance;
- Cell delay variation tolerance (CDVT);
- QoS class and/or parameters.

Peak Cell Rate

PCR in cells/second is a traffic parameter that specifies an upper boundary on the traffic that can be submitted on an ATM connection.

Sustainable Cell Rate

The upper boundary on the conforming average rate of an ATM connection is called the SCR in cells/second. SCR is always less than or equal to PCR.

Maximum Burst Size

MBS is the maximum number of cells that can be sent at the peak rate.

Burst Tolerance

Burst tolerance is conveyed through the MBS in number of cells.

Cell Delay Variation Tolerance

CDVT is described in the QoS section.

Traffic Descriptor

The ATM traffic descriptor is the list of traffic parameters that is used to define the traffic characteristics of an ATM connection (unidirectional).

Source Traffic Descriptor

The source traffic descriptor is a list of the traffic parameters for an ATM connection that is used during the connection setup for a particular source. The source traffic descriptor may vary from connection to connection.

Connection Traffic Descriptor

The traffic descriptors is a set of traffic parameters that specify the traffic characteristics of an ATM connection. It contains the necessary information for conformance-testing of the ATM cells of an ATM connection. CDVT and the conformance definition specify the conforming cells of the ATM connection. Conformance testing is performed using a preestablished algorithm that either discards or tags nonconforming cells. Connection admission control procedures use these descriptors to allocate network resources (e.g., bandwidth) efficiently.

Conformance Definition

The generic cell rate algorithm (GCRA) is used to specify the conforming cells of an ATM connection at the UNI or NNI. The first cell of the connection initiates the algorithm, and subsequent cells are either conforming or nonconforming. The GCRA is discussed later in this chapter.

Traffic Management

ATM networks have essentially three traffic-related objectives:

1. Handle different traffic types with different QoS requirements;
2. Ensure fairness among different customers, subscribing services for different traffic types;

3. Optimize network resource (e.g., numbers of switches, links, and bandwidths) use without service degradation.

To meet these network objectives, several traffic management techniques are used that can be broadly classified into the following three categories:

1. *Congestion avoidance and ensuring fairness*: These include negotiating a traffic contract and enforcing it, so that violating traffic can be discarded and not be able to adversely affect conforming traffic. It includes traffic shaping, where traffic transmitted from a network node is controlled based on preestablished criteria;
2. *Traffic control and congestion control*: These include traffic priority control, where traffic with different priority or QoS class is treated differently by the network, such as selective cell discard and congestion notification;
3. *Traffic engineering*: It allows network resource optimization by modeling traffic sources and switch performance.

Refer to the ATM Forum document AF-TM-0056.000 and ITU-T document I.371 for more information.

Traffic Control and Congestion Control

Traffic control and congestion control functions are a set of actions taken by the network in all relevant network elements. Under normal conditions, (e.g., when there are no network failures), the traffic control functions are intended to avoid network congestion. Nevertheless, congestion may occur. This may be brought on by unpredictable statistical fluctuations of traffic flows or by network failures. Additional functions, referred to as congestion control functions, are intended to minimize the intensity, spread, and duration of network congestion and to manage the effects of congestion for different classes of traffic differently.

A range of traffic and congestion control functions are used in the B-ISDN to maintain the QoS of ATM connections as shown in Figure 14.2.

Various traffic and congestion control schemes proposed for ATM networks can be categorized into the following two groups:

1. Preventive control;
2. Reactive control.

Preventive Control

Preventive control schemes attempt to prevent congestion from ever occurring. The following mechanisms can be used to minimize network congestion:

- Resource allocation: Allocation of the trunk bandwidth and the buffer size based on connection traffic descriptors and optimal oversubscription;

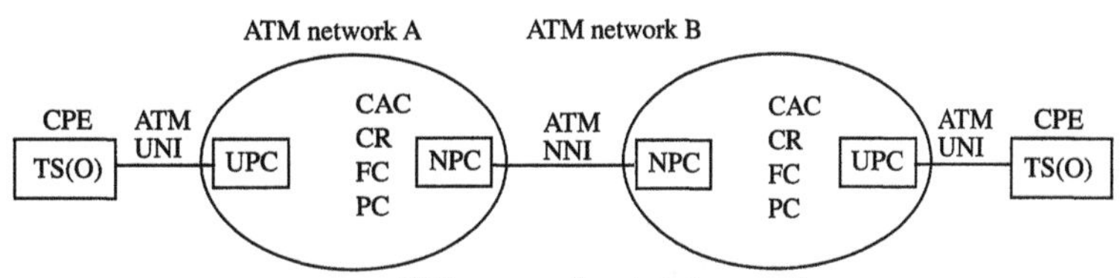

CAC: connection admission control
CR: call routing
FC: feedback control
PC: priority control
NNI: network-to-node interface
UPC: usage parameter control
NPC: network parameter control
TS(O): traffic shaping (optional)
CPE: customer premises equipment
UNI: user-to-network interface

FIGURE 14.2. Traffic and congestion control functions in ATM networks. (*After:* I-371, ITU, 1995.)

- Usage parameter control (UPC): Ability to tag or discard cells that are not conforming to the traffic contract;
- Connection admission control (CAC): Book new connections for their maximal capacity;
- Traffic engineering: Provision trunk bandwidth and switch buffers based on traffic analysis, traffic modeling, and long-term projection of traffic growth.

Reactive Control

Reactive control schemes attempt to minimize the effect of congestion when it does occur. The following mechanisms can be used to avoid severe congestion:

- UPC: Cell tagging, cell discarding;
- CAC: Overbook new connections beyond their maximum capacities;
- Flow control: Control the rate of cell transfer using explicit forward congestion indication (EFCI) (e.g., end-user notification), explicit rate marking;
- Connection blocking: Block new connection request;
- Disconnect calls: Disconnect active calls.

Each of the two groups mentioned above is applicable at different time scales. The time scale over which a particular control is applicable, 7 is generally very important. The minimal time for traffic or congestion control to take effect is a cell time (generally few microseconds). Figure 14.3 shows the traffic and congestion control options and their relative time scales.

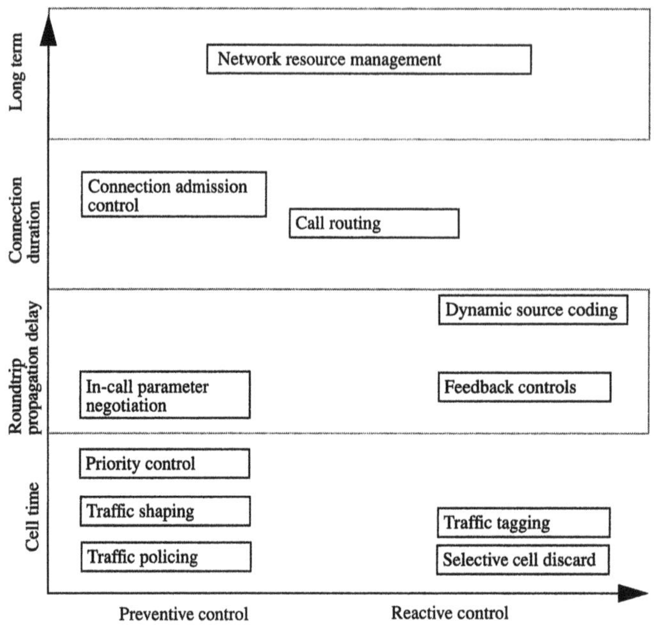

FIGURE 14.3. Traffic and congestion control and their time relative scale. (*After:* R. Handel, M. N. Huber, S. Schroder, *ATM Networks, Concepts, Protocols, Applications*, Addison-Wesley, 1994.)

The traffic control mechanisms used at the cell level are limited to preventive measures, such as traffic policing and traffic shaping, which are based on preestablished parameters.

The roundtrip propagation delay (e.g., generally in milliseconds) is applicable for reactive control mechanisms that are based on congestion detection followed by some action at the source or upstream node.

The connection duration (which can be from a few seconds to many minutes) is applicable to traffic control mechanisms, such as connection admission control and call routing.

The long term is applicable to traffic control mechanisms, such as resource provisioning.

The descriptions of the following traffic control and congestion control functions are given below:

Traffic Control Functions:

- Network resource management;
- Connection admission control;

- Usage parameter control;
- Priority control;
- Generic flow control;
- Selective cell discarding;
- Traffic shaping;
- Feedback control;
- Explicit forward congestion indication.

Congestion Control Functions:

- Selective cell discarding;
- Explicit forward congestion indication.

Additional Traffic Control Functions:

Other useful techniques that can be used are:

- Connection admission control that reacts to and takes into account the measured load of the network;
- Variation of usage monitored parameters by the network, such as a reduction of the peak rate available to the user;
- Traffic control techniques such as rerouting and connection release;
- Fast resource management.

Network Resource Management (NRM) and Provisioning

Network resource management (NRM) controls how network resources such as transmission links and network nodes (e.g., switches, multiplexers) are provisioned. The NRM objective is to optimize network use and at the same time reduce network operation costs. It is desirable to separate traffic flows according to the service characteristics. Therefore, VPCs are important components of resource management in ATM networks. VPCs can be used in the following manner:

- To efficiently route many virtual connections with same destinations;
- Aggregate virtual connections with the same QoS requirements, so that intermediate network nodes need to process traffic only at the virtual path level;
- Aggregate virtual connections such that the UPC can be applied to the traffic aggregate.

VPCs can also play an important role in reducing network operation cost. The processing required to establish individual VCCs can be reduced. Individual VCCs can be established by making simple connections at the nodes where the VPCs terminate. Different VCCs routed through the same VPCs in the same sequence will experience similar expected network performance along the way. These performance metrics are cell loss ratio, cell delay variation, and cell transfer delay. When

VCCs within a VPC require a range of QoS, the VPC objective is to accommodate the most demanding VCC being carried.

The intermediate nodes have no knowledge of the QoS of the individual VCCs within the VPC. The combined peak of all of the VC links could exceed the VPC capacity only if all of the VC links within the VPC can tolerate the QoS that would result from this statistical multiplexing.

When statistical multiplexing of VC links is applied by the network operator, different VPCs may be used to separate traffic with different QoS classes. To facilitate a full range of QoS between the origination and destination points, several VPCs may be necessary.

Connection Admission Control

The connection admission control (CAC) is a set of actions taken by the network to determine whether a VCC or a VPC can be accepted or should be rejected. This occurs during the call setup phase or during the call renegotiation phase.

In the ATM network, a connection request is accepted or rejected based on the availability of resources to establish a connection through the whole network and maintain its required QoS. This must be accomplished without compromising the QoS of already-existing connections. These decisions are based on CAC and apply to renegotiated connection parameters within a given call as well.

The CAC function has to derive the following information from each traffic contract (e.g., values of parameters in the source traffic descriptors and the requested QoS class) for each connection request:

- Peak cell rate (PCR);
- The value of the cell delay variance tolerance (CDVT);
- Sustainable cell rate (SCR);
- Maximum burst size (MBS);
- The requested conformance definition;
- The requested QoS class.

The CAC uses this acquired information and the network operator definition of a compliant connection to decide:

- Whether to accept the connection or reject it;
- The routing and allocation of network resources.

Figure 14.4 shows a simple block diagram using CAC.

Different strategies of network resource allocation may be used for CLP=0 and CLP=1 traffic. If tagging is not requested, the network may still tag nonconforming cells, so the CAC may accept a cell even though tagging was not requested. When performing CAC, information of the measured network load may be used. This will allow the network operator to gain a higher network use without compromising the network performance objectives.

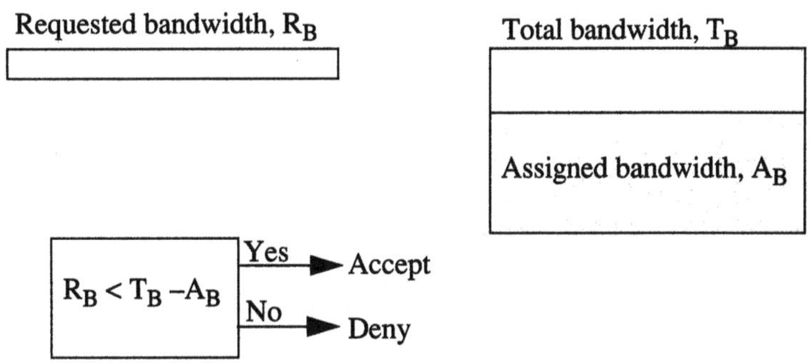

FIGURE 14.4. Connection admission control.

Usage Parameter Control

UPC Functions

Usage parameter control is the set of actions taken by the network to monitor and control traffic offered by the source for a specific ATM connection at the user interface. The main purpose of UPC is to protect the network against accidental as well as intentional violations of negotiated parameters that can affect the QoS of already-established connections.

UPC applies to both user and OAM flows. The task of monitoring performed by the UPC for VCCs and VPCs is accomplished by the following steps respectively:

- Checking the validity of VPI and VCI; that is, whether or not their values are associated with an active VCC, and monitoring the traffic that is entering the network to ensure no traffic parameter is being violated;
- Checking the validity of the VPI; that is, whether or not the VPI value is associated with an active VPC, and monitoring the traffic that is entering the network to ensure no traffic parameter is being violated.

UPC Requirements

Some of the features of the UPC algorithm are:

- The ability to detect noncompliant traffic;
- Selectivity over the range of parameters checked. The algorithm could determine whether the user's behavior is within acceptable boundaries;
- Rapid response to parameter violations;
- Simplicity of implementation.

There are two performance parameters that must be considered when assessing the performance of UPC mechanisms:

- *Response time*: The time to detect a noncompliant situation on a VPC or VCC under the given reference conditions;
- *Transparency*: Under given reference conditions, the accuracy with which the UPC begins the appropriate control actions on a noncompliant connection and avoids any inappropriate action with a compliant connection.

The impact of UPC on cell delay is also considered. Any cell delay or CDV that is introduced by the UPC must be taken into account as part of the delay and delay variation allocated to the network.

UPC Location

Usage parameter control is performed on VCCs or VPCs at the first point where VP or VC is terminated.

Traffic Parameters Controlling the UPC

The source traffic descriptor contains the traffic parameters that control UPC.

UPC Actions

The function of the UPC is to control traffic offered by an ATM connection and to ensure that it conforms with the negotiated traffic contract. Some of the functions of the UPC at cell level are:

- Cell passing of conforming cells;
- Cell tagging (based on traffic contract) of nonconforming cells. Cell tagging operates only on CLP=0 cells by overwriting the CLP bit to 1;
- Cell discarding of nonconforming cells.

Relationship Between UPC and CLP

CLP bit in the ATM cell header allows added flexibility in how individual cells are treated by the UPC. Depending on the traffic contract, the user can transmit cells with CLP=0 or CLP=1, and the UPC can treat them the same way. The UPC can also monitor only CLP=0 cells and tag them when they are violating the traffic contract. No resource is committed for CLP=1 cells in this case, and they are discarded if congestion occurs.

Relationship Between UPC and OAM

OAM cell flows across the UNI are also subject to policing. The network may require some knowledge from the user of the traffic parameters, such as the peak cell rate and knowledge of some clumping tolerance for the OAM traffic of an ATM connection. Whether or not the user explicitly or implicitly specifies the OAM cell stream, the network may police the OAM cell flows separately from user data cellstreams or may police OAM cell flows together with user data cellstreams. The traffic parameters for OAM cell flow across the UNI may be explicitly specified at the time of subscription or implicitly by a default rule.

Priority Control

Priority control allows the service provider to establish different priorities for different services based on the QoS requirements. This can be accomplished by what is also called *priority queueing*, *service scheduling*, or *fair queueing*. Basically, multiple queues are implemented in the ATM switches, such that traffic on certain connections that have higher priority can jump ahead of those with lower priority.

Generic Flow Control (GFC)

Four bits of the ATM cell header at the UNI are dedicated for generic flow control (GFC). ATM terminal may have either a single traffic type or multiple traffic types. A terminal with multiple traffic types may have multiple prioritized queues as discussed earlier. This field may carry messages, such as start, cells for queue 1, stop, cells for queue 1, or start, cells for queue 2, stop, cells for queue 2. How to use these bits efficiently is still under discussion.

Selective Cell Discard

A cell loss priority (CLP) bit may be used by users or network to prioritize individual cells. Elements of the network may selectively discard cells of the CLP=1 while still meeting the network performance objectives on the CLP=0 flows. A congested network element may selectively discard cells explicitly identified as belonging to a noncompliant ATM connection or those cells with CLP=1. This measure is taken to protect the CLP=0 flows as long as possible.

The CLP bit can be set by the network as a result of the policing function when the cell is tagged as a nonconforming cell. This bit is also set by users to prioritize their cells before they are transmitted to the network. For example, an MPEG-2–encoded video source may put higher priority on the cells containing the initial frame of a scene and lower priority on the cells containing the intermediate or predictive frames that are not as critical for decoding and quality of the resulting video.

Two most common selective cell discard mechanisms are:

1. Push out;
2. Threshold.

Push Out

In the push out mechanism, a higher priority cell can remove a lower priority cell from the queue, if the queue is full when it arrives.

Threshold

In the threshold mechanism, a buffer occupancy threshold is used to regulate the admission of low-priority cells into the queue. If the threshold exceeds a certain point, all low-priority cells are discarded until the queue size falls below the threshold. High-priority cells are continued to be admitted as long as there is space in the queue.

Traffic Shaping

Traffic shaping is the mechanism by which the desired characteristics of the stream of cells into a VCC or VPC is attained at the source ATM endpoint. It ensures that they do not transmit cells at higher peak or average rates than those they have committed to. Of course, it is difficult to mandate that all nodes do, in fact, implement traffic shaping internally, because only their output cellstreams can be observed.

Traffic shaping in a private ATM switch is the mechanism by which the traffic characteristics of a stream of cells on a VCC or VPC are altered to achieve a desired modification of the traffic characteristics. Traffic shaping always maintains the integrity of the cell sequence on the ATM connection. Traffic shaping examples:

- Peak cell rate reduction;
- Burst length limiting;
- Reduction of cell clumping due to CDV by appropriately spacing the cells in time.

An ATM end station may choose to shape to the negotiated peak cell rate for the aggregate cellstream of (CLP=0 + CLP=1) cells and may choose not to shape to the negotiated peak cell rate for the CLP=0 cellstream and instead to allow the network UPC mechanism to tag CLP=0 cells if they violate the conforming parameters.

Some of the following mechanisms can be used to shape traffic:

- *Buffering*: Buffer cells to ensure that the traffic contract will not be violated (GCRA is used to determine when to allow cells to enter the network);
- *Spacing*: Police cells from all VCs as one and maintain the intercell spacing as they enter the network. Some cells may be discarded because of input burst that cannot be spaced out within the specified burst tolerance;
- *Scheduling*: Queue cells from multiple VCs and schedule their entry to the network to ensure that the traffic contract will not be violated;
- *Peak cell rate reduction*: Cell entry into the network at peak rate less than that in the traffic contract;
- *Source rate limitation*: Limit actual source rate in some way;
- *Burst length limiting*: Cell entry into the network at MBS less than that in the traffic contract;

- *Priority queuing*: Multiple queues are implemented in the ATM switches, such that traffic on certain connections that have higher priority can jump ahead of those with lower priority;
- *Framing*: Use a frame structure to schedule cell entry into the network.

Feedback Control

The preventive control mechanisms described above can minimize the number of occurrences of congestion, but they cannot fully be eliminated because of the statistical nature of the traffic.

The preventive mechanisms all depend on the source's ability to characterize their traffic accurately and declare the values of the traffic parameters within which they will behave. Accurately characterizing sources is a difficult process and involves uncertainties, however, in addition, some sources may not be capable or willing to characterize their traffic at all. For example, many applications have the ability to reduce their information transfer rate if the network requires them to do so; likewise, they may wish to increase their information transfer rate if there is extra bandwidth within the network. Therefore, in addition to preventive controls, reactive controls or feedback controls are required to monitor the congestion level in the network. The network notifies the sources when congestion is detected and the sources will take action based on the congestion information. The main objective of the feedback control is to prevent momentary periods of traffic surge that may cause cell losses. The following two feedback control methods are discussed below:

- End-user notification;
- Adaptive rate control.

End-User Notification

End users are notified after the network experiences congestion. The idea is that end users will take the necessary actions after they have been notified to avoid worsening the congestion. The following two mechanisms are used to notify end users:

1. Explicit forward congestion indication (EFCI);
2. Backward explicit congestion notification (BECN).

Explicit Forward Congestion Indication (EFCI)

The EFCI is a congestion notification mechanism that the ATM layer service user can employ to improve the utility that can be derived from the ATM layer. The use of this mechanism by the CPE is optional, and the network operators do not rely on this mechanism to control congestion.

A network element in a congested state or in an impending congested state may set an EFCI in the ATM cell header (the center bit of the 3-bit payload type field) so that it may be examined by the destination CPE. The CPE may use this indication to implement protocols to lower the cell rate of the connection during the congestion or to ward off the impending congestion. A network element that is not in a congested state nor an impending congested state does not modify this bit. An impending congested state is reached when a network resource is operating near its engineered capacity.

Congestion detection can be based on buffer occupancy threshold or some other methods. The intermediate nodes along the path to the end node (or end user) cannot modify this bit. The use for this notification for actions to be taken by the end node (or user) has not been standardized. Because of large propagation delays relative to transmission times of cells, by the time a marked cell arrives at a receiver, it is possible that the congested node that marked the cell is no longer congested. The receivers should not react to EFCI very quickly. Instead, statistics should be collected to accurately determine whether there is momentary or sustained congestion along the path. If sustained congestion is detected, then the receiver should send a notification back to the source to slow down.

Backward Explicit Congestion Notification

In this mechanism, each node in the network monitors the congestion status. When congestion is detected, a special cell, called the resource management (RM) cell, is sent upstream to the source nodes of all connections that pass through the congested node. As in EFCI, congestion detection can be based on buffer occupancy of the trunks. This method minimizes the time it takes to notify a source about network congestion. Because a special cell is used for this purpose, the information fields can be used to include a variety of information. The sources can react most effectively to congestion along their paths.

Adaptive Rate Control

In adaptive rate control, the source varies its transmit rate based on the congestion status feedback received from the network. The basic idea is to allow the sources to start from an initial transmit rate of zero and gradually increase their rate. The rate is roundtrip time (RTT). With every RTT, the network provides feedback on whether the rate can be increased or decreased. As long as congestion is not detected, each source is allowed to increase its transmit rate by an additive increase (e.g., five cells per RTT). When congestion is detected, feedback is provided to the sources indicating that they should decrease their transmit rate by a multiplicative (drastic) decease (e.g., multiplicative factor = 0.5). When congestion has passed, the process is repeated.

Available Bit Rate

Available bit rate (ABR) is defined as an ATM layer service that can request available bandwidth during connection setup. The sources adapt to the changing network conditions. When bandwidth is available, they use as much of it as possible. When it is not available, they can throttle back.

An example of EFCI and BECN is shown in Figure 14.5.

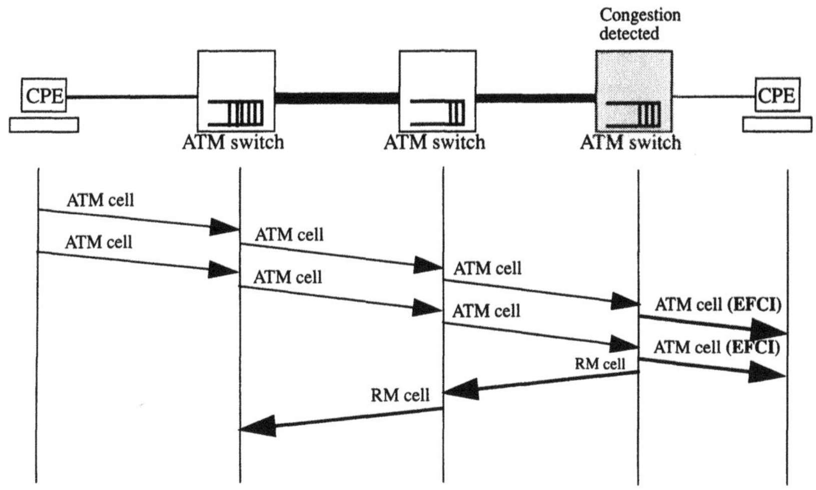

FIGURE 14.5. Example of EFCI and BECN.

Resource Management

The following two critical resources should be managed by the ATM network:

1. Buffer space;
2. Trunk bandwidth.

The trunk bandwidth can be managed through the use of virtual paths (VP). Because a virtual path is constructed with multiple VCCs, management is easier than managing the individual VCCs.

Fast Resource Management

Fast resource management (FRM) cells are used for managing bandwidth and the buffer resources dynamically. The payload type (PT) field of the ATM cell header is coded with 110 to denote an FRM cell. Figure 14.6 shows an FRM cell format.

The fast bandwidth reservation basically involves reserving trunk bandwidth at each node of the intended connection. Two types of bandwidth reservation can be requested. These are immediate unguaranteed and delayed guaranteed.

Fast buffer reservation is used to reserve buffer space in the intermediate nodes of the connection for the temporary surge of ATM cells.

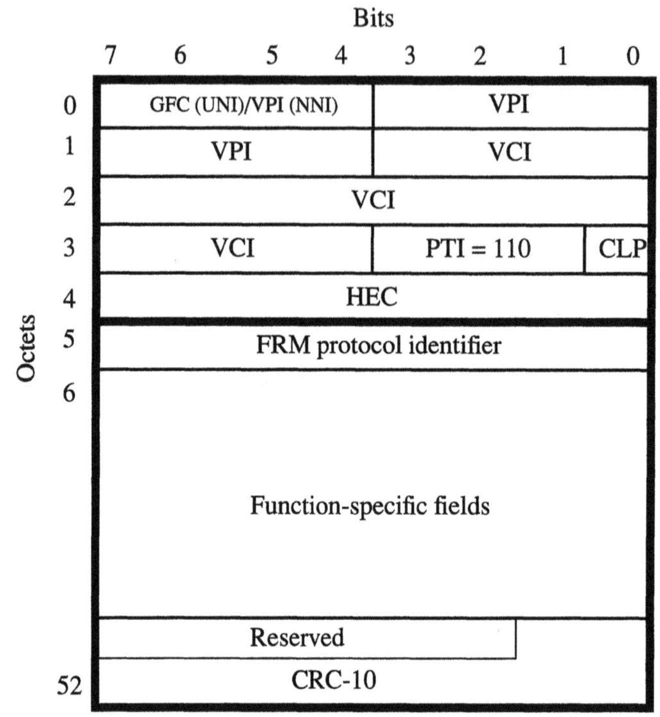

FIGURE 14.6. FRM cell format.

FRM message flow for bandwidth or buffer reservation is shown in Figure 14.7.

Generic Cell Rate Algorithm

The GCRA is a virtual scheduling algorithm or a continuous state leaky bucket algorithm defined in ATM standards. The GCRA is used to define the PCR and CDV tolerance, sustainable cell rate, and burst tolerance.

GCRA is also used to provide the formal definition of traffic conformance to the traffic contract. For each arriving cell, the GCRA determines whether the cell is conforming to the traffic contract of an ATM connection.

For any sequence of cell arrival times $\{t_a(k), k \geq 1\}$, both algorithms determine the same cells to be conforming or non-conforming. The GCRA uses two parameters: the increment I and the limit L. These parameters are also denoted by T and tau, respectively.

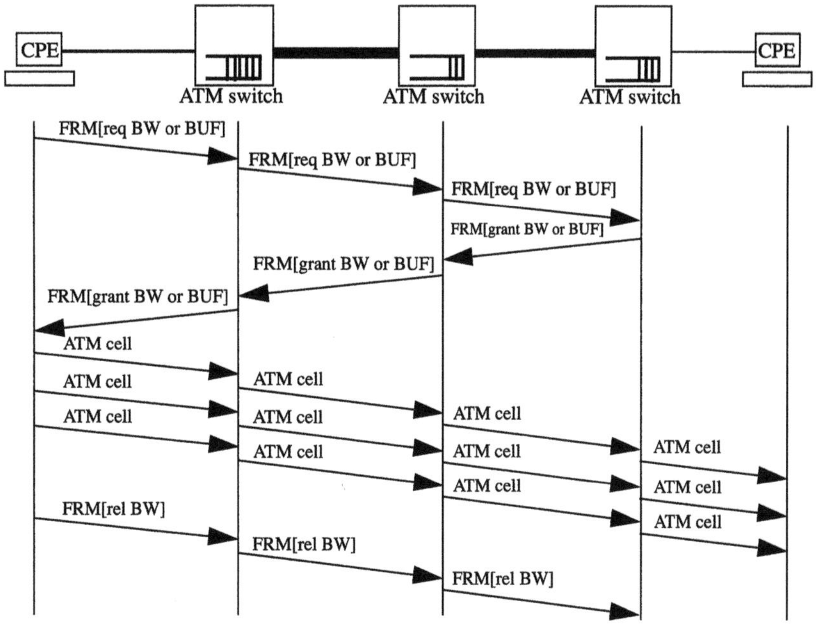

FIGURE 14.7. FRM message flow.

The GCRA (*I*,*L*) means the generic cell rate algorithm with the value of the increment set equal to *I* and the value of the limit parameter set equal to *L*.

Virtual Scheduling Algorithm

At the arrival of the first cell, the theoretical arrival time *TAT* is initialized to the current time, $\{t_a(1) + I\}$. This is the predicted arrival time for the second cell. If the second cell arrives later than *TAT*, that means $TAT < t_a(2)$, then the *TAT* is recalculated to $\{t_a(2) + I\}$.

For subsequent cells:

- If the arrival time of the kth cell, $t_a(k)$, is actually after the current value of the *TAT* then the cell is conforming and *TAT* is updated to $\{t_a(k) + I\}$;
- If the arrival time of the kth cell, $t_a(k)$, is actually before the current value of the *TAT* and greater than or equal to *TAT* + *L*, then again the cell is conforming and *TAT* is updated to $\{t_a(k) + I\}$;
- If the arrival time of the kth cell, $t_a(k)$, is actually before the current value of the *TAT* + *L*, then the cell is not conforming and *TAT* is not updated and it remains $\{t_a(k-1) + I\}$.

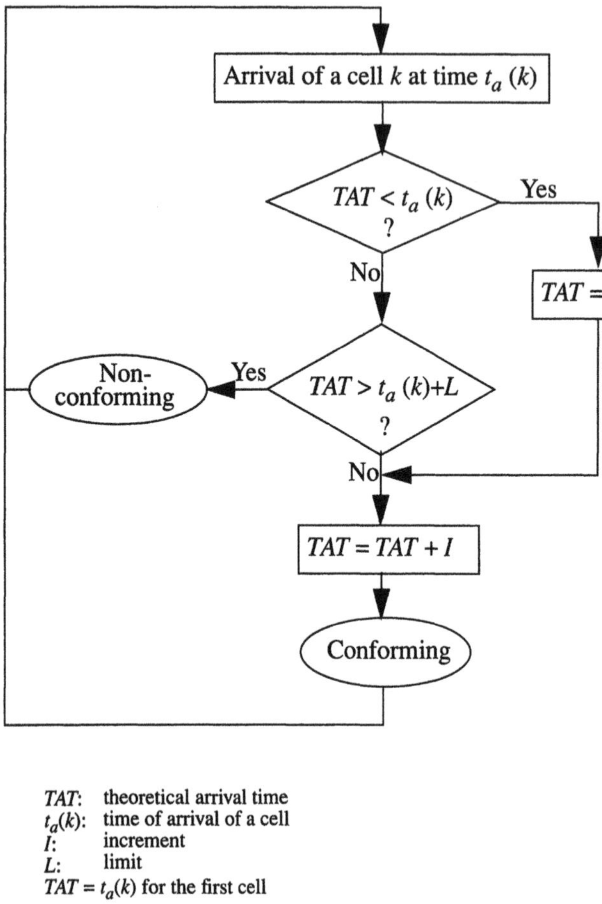

FIGURE 14.8. Virtual scheduling algorithm. (*After:* I-371, ITU, 1995.)

Figure 14.8 shows the virtual scheduling algorithm flow diagram.

Continuous-State Leaky Bucket Algorithm

Leaky Bucket Concept

The continuous state leaky bucket algorithm can be viewed as a bucket with a hole in the bottom. Each time a cell arrives, a fixed quantity of water is added to the bucket. The water drains out constantly. A cell is conforming (can be admitted) as long as the bucket is not full. A cell is nonconforming when it causes the bucket to overflow.

Let

- X_{NEW} = New contents of the bucket;
- X_{OLD} = Old contents of the bucket;
- E_{LT} = Elapsed time since arrival of the last cell;
- D_R = Drain rate;
- D_L = Bucket limit.

Then

$$X_{NEW} = X_{OLD} - E_{LT} \times D_R + 1$$

We can convert the above equation in time units by dividing both sides of the equation by D_R

$$\frac{X_{NEW}}{D_R} = \frac{X_{OLD}}{D_R} - \frac{E_{LT} \times D_R}{D_R} + \frac{1}{D_R}$$

Therefore

$$X_{NEWT} = X_{OLDT} - E_{LT} + I$$

Where

$$I = \frac{1}{D_R}$$

Figure 14.9 shows the leaky bucket block diagram.

Standard Leaky Bucket Algorithm

At the arrival of the first cell, the content of the bucket, *X*, is set to 0 and the last conformance time *LCT* is initialized to the current time, t_a (1).

For subsequent cells:

- At the arrival time of the kth cell, t_a (*k*), the value of *X'* is calculated, which equals the content of the bucket, *X*, after the arrival of the last conforming cell minus the amount the bucket has leaked since that arrival. Content of the bucket is set to 0 if it becomes a negative number. Then, if *X'* is less than or equal to the limit value *L*, then the cell is conforming and the bucket content *X* is updated to {*X'* + *I*} for the current cell, and the *LCT* is also updated to the current time t_a (*k*);
- If *X'* is greater than the limit value *L*, then the cell is nonconforming and the values of *X* and *LCT* are not updated.

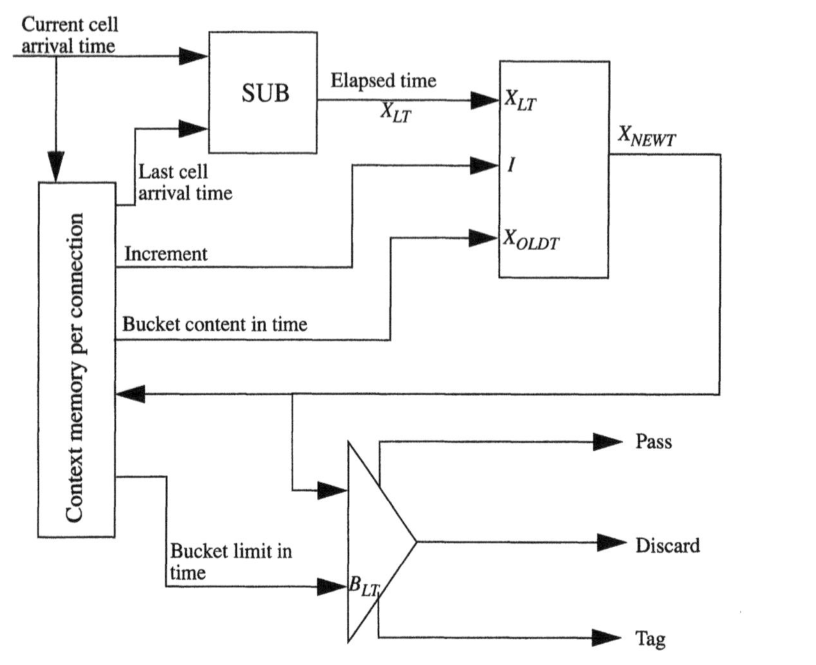

FIGURE 14.9. Leaky bucket block diagram.

Figure 14.10 shows the continuous-state leaky bucket algorithm flow diagram.

GCRA Examples

The virtual scheduling algorithm or the continuous-state leaky bucket algorithm is equivalent and for any sequence of cell arrival time both algorithms determine the same cells to be conforming or nonconforming.

Figure 14.11 shows the ATM cellstream.

Virtual Scheduling Algorithm Example

A virtual scheduling algorithm GCRA (I,L) is defined for an ATM connection, where, *TAT* = Theoretical Arrival Time, $T_a(k)$ = cell arrival time, I = Increment, and L = Limit.

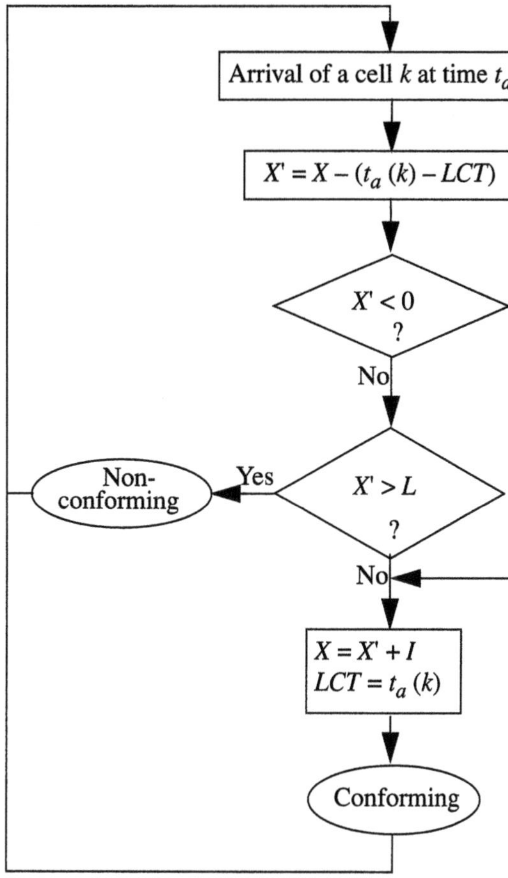

LCT: last compliance time
$t_a(k)$: time of arrival of a cell
X: value of the leaky bucket counter
X': auxiliary variable
I: increment
L: limit
$LCT = t_a(k)$, $X = 0$ for the first cell

FIGURE 14.10. Continuous-state leaky bucket algorithm. (*After:* I-371, ITU, 1995.)

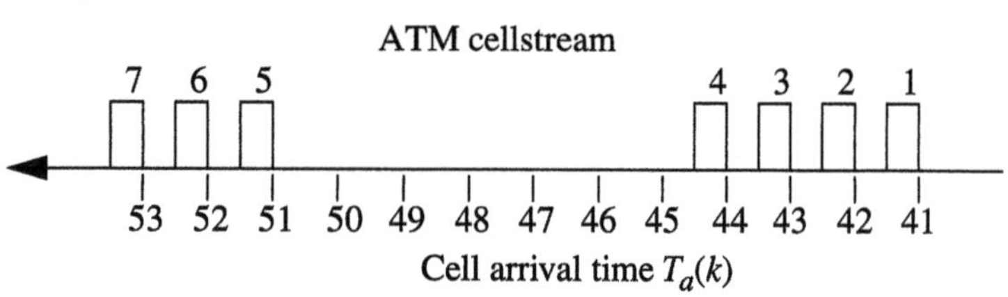

FIGURE 14.11. ATM cellstream.

The following table is constructed using GCRA (3,2):

	ATM cell						
	1	2	3	4	5	6	7
$T_a(k)$	41	42	43	44	51	52	53
TAT	41	45	48	48	54	57	57
pass/discard	pass	pass	pass	disc	pass	pass	disc

The following table is constructed using GCRA (5,4):

	ATM cell						
	1	2	3	4	5	6	7
$T_a(k)$	41	42	43	44	51	52	53
TAT	41	47	52	52	57	57	62
pass/discard	pass	pass	pass	disc	pass	disc	pass

Continuous-State Leaky Bucket Algorithm Example

A continuous-state leaky bucket algorithm GCRA (*I*,*L*) is defined for an ATM connection, where, $T_a(k)$ = cell arrival time, *I* = Increment, *L* = Limit, *X* = leaky bucket counter value, *X'* = auxiliary variable, and *LCT* = Last Compliance Time.

The following table is constructed using GCRA (3,2):

	ATM cell						
	1	2	3	4	5	6	7
$T_a(k)$	41	42	43	44	51	52	53
X'	0	0	2	4	0	2	4
X	0	3	5	5	3	5	5
LCT	41	42	43	43	51	52	52
pass/discard	pass	pass	pass	disc	pass	pass	disc

The following table is constructed using GCRA (5,4):

	ATM cell						
	1	2	3	4	5	6	7
T_a(k)	41	42	43	44	51	52	53
X'	0	0	4	8	1	5	4
X	0	5	9	9	6	6	9
LCT	41	42	43	43	51	51	53
pass/discard	pass	pass	pass	disc	pass	disc	pass

Traffic Engineering

To achieve better performance from the ATM network, traffic sources and the switch performance should be modeled properly. Because traffic engineering is a subject by itself, it is not covered in this book.

CHAPTER 15

ATM Network Management

ATM Network Management

The term ATM network management refers to the means by which the central administrative authority monitors the behavior of the network and controls the network elements. There are well-established standards for the management of public switching systems. The public ATM networks will likely have to meet those standards.

ATM network management framework is defined for the following three areas:

1. Interface management;
2. Layer management;
3. Network management.

Interface Management

Interface management collects information and configures various ATM interfaces. The ATM Forum has defined the interim local management interface (ILMI) in the UNI interface. The ATM DXI and LAN emulations also fall in this category. The ILMI is described later in this chapter.

Layer Management

Layer management deals with segment or end-to-end virtual path and virtual channel management through OAM cells. The layer management functions (e.g., fault management, performance management, activation/deactivation, and system management) are described in Chapter 9.

Network Management

Network management deals with the multiple ATM switches that monitor and control ATM devices in the total ATM network. The ATM network management system (NMS) includes manager, agent, management information base (MIB), and protocol.

The network manager is a computer that translates human commands to monitor and control the devices in the ATM network.

The agents, also known as network management entities (NME), are software for doing network management functions. Each agent collects data from the network and stores them locally for future references. Agents also respond to the commands sent by the manager.

There are multiple objects for defining different aspects of a managed agent. A manager can alter the network setting by changing the value of the corresponding object. The collection of these objects is known as MIB.

The manager and agents communicate with one another using a network management protocol. The protocols used to communicate between the manager and the management agents vary between the public and private networks. The public networks use the international organization for standardization (ISO) common management and information protocol (CMIP), that is more common in the public network. The private networks use the IETF simple network management protocol (SNMP).

ATM Network Management Functions

ATM network management covers the following functions, described below:

Configuration Management

Configuration management configures network elements (NE), UNIs, VPCs, VCCs, and cross-connect VP/VC links. Some of the configuration management functions are as follows:

- Setting up and tearing down connections;
- Setting the number of connections supported by an interface;
- Setting the number of PVCs;
- Setting the number of VPI/VCI bits supported for address translation;
- Retrieving connection status.

Fault Management

Fault management reports NE failure, module failure, ATM link or connection failure. It initiates ATM-OAM cell loopback tests and physical layer (e.g., SONET, DS-3 level) loopback tests to locate network faults. The network is reconfigured quickly to minimize the impact of a failure. Some fault management functions are as follows:

- Locating the faulty device by displaying physical alarms (e.g., lamps) and message display or printout;
- Sending alarm messages to other systems in the network affected by the fault condition;
- Reconfiguring the network to isolate the faulty equipment;
- Restoring the network.

Performance Management

Performance management monitors physical layer (e.g., SONET, DS-3 level) performance, TC and ATM protocol, UPC and NPC violations, and NE congestion and use. Some performance management functions are as follows:

- Checking the QoS requirement;
- Collecting statistics for alarms;
- Preparing the statistics for traffic contract violation;
- Supporting ATM layer performance management functions (e.g., OAM flows).

Security Management

Security management identifies and authenticates (e.g., login/password) users for access control. It protects information with integrity and confidentially during transfer to another location. Some security management functions are as follows:

- Identification and authentication of information source before retrieval;
- Verification of information for integrity;
- Maintenance of confidentiality of the information received;
- Acknowledgement sent for every transfer.

Accounting Management

Accounting management logs the usage of the network resources by users and determines the cost associated with it. Some of the accounting management functions are as follows:

- Recording of all items relating to a connection (e.g., duration of call, QoS parameters used, number of successfully transmitted cells, errored cells);
- Preparing of the billing information;
- Transmitting of the billing information to a processing center.

ATM Network Management Protocol

The manager communicates with several agents to control the overall network management. The objects belonging to each agent are controlled (e.g., updated) as directed by its manager, and the results are reported back to its manager.

Two standard protocols used to communicate between the manager and the management agents are:

1. Simple network management protocol (SNMP);
2. Common management and information protocol (CMIP).

Simple Network Management Protocol

SNMP is defined in IETF RFC 1157 and is widely used in private networks (e.g., TCP/IP networks). SNMP supports only two basic services: fetching and storing variables. This protocol consists of five types of operations as shown in Table 15.1. SNMP messages are normally 484 octets long. Longer-than-484-octet messages can be negotiated.

Common Management and Information Protocol

CMIP is defined by International Organization for Standardization (ISO) for public networks (e.g., service provider networks). It is used for overall network management and supports powerful and complex commands.

ATM Network Management Reference Model

ATM network management reference model defined by ATM Forum has five management interfaces. These are referred to as M1 through M5 as follows:

- M1: Management of ATM user devices;
- M2: Management of private ATM networks;
- M3: Management of two management systems, one in a public ATM network and the other in a private ATM network, to communicate with each other;
- M4: Management of public ATM networks;
- M5: Management of two management systems, both in a public ATM network, to communicate with each other.

TABLE 15.1. SNMP Operations

Operation	Function
Get	Retrieve specific management information
Get-next	Retrieve in steps
Get-response	Replies to retrieve command
Set	Change management information
Trap	Report exceptions

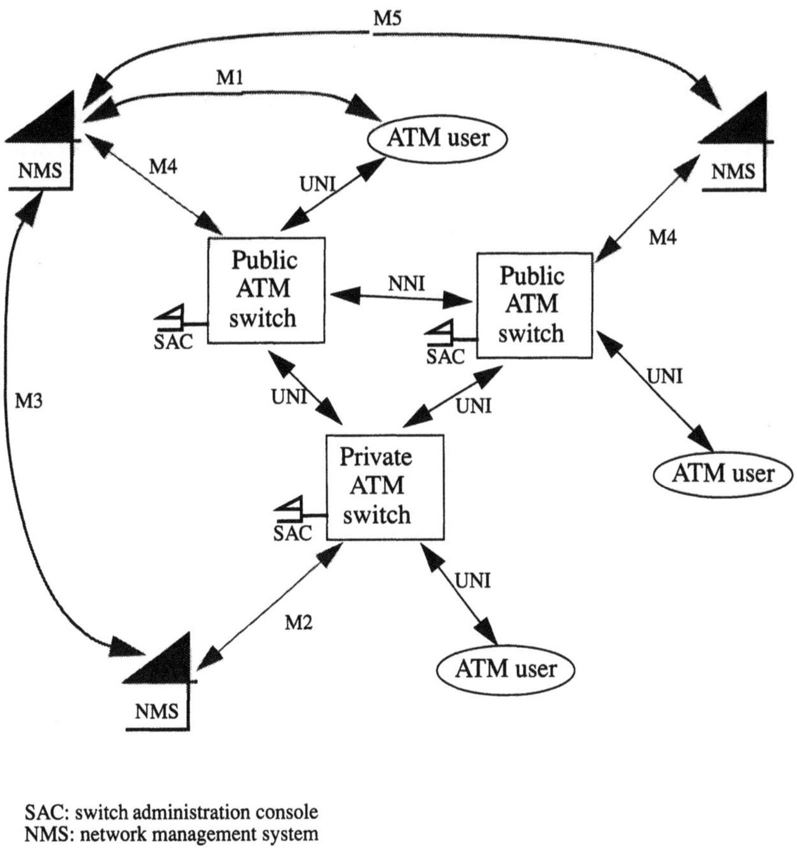

FIGURE 15.1. ATM network management interfaces.

The ATM network management interfaces are shown in the Figure 15.1.

Interim Local Management Interface

The ATM Forum has concentrated on devising a management framework for private ATM implementation. This allows two management entities in the nodes that are connected by a UNI to manage that UNI through a symmetric, peer-to-peer protocol. The interim local management interface (ILMI) is a preliminary (interim) standard that may be replaced by ITU-T. ILMI provides any ATM user device with status and configuration information about VP/VC connections at the UNI and network management information.

The ILMI standard is based on the SNMP which is now the standard for multivendor network management. It defines an MIB for UNI, together with a protocol stack—SNMP running directly over AAL-5 (or AAL-3/4 optional) and a reserved VPI/VCI. This protocol stack is used by peer entities to manage a single UNI. An ATM switch would support multiple management entities and associated MIBs, one for each UNI. The ILMI is only one management element of an entire network, which might consist of public and private switches, as well as UNI.

The ILMI is integrated with the overall management model for an ATM device as shown in Figure 15.2.

It is likely that both public and private ATM switches will support their own management agents and MIBs. These MIBs would contain information on the operation of the whole switch, not just the UNIs. The management agents will be monitored and controlled by a central NMS in both public and private sectors.

The protocols used to communicate between the NMS and the management agents could vary between the public and private networks. The public networks use the OSI CMIP. The private networks use SNMP. Communication between the public and private NMSs, known as customer network management (CNM), provides users with limited access to information about control over their public UNI.

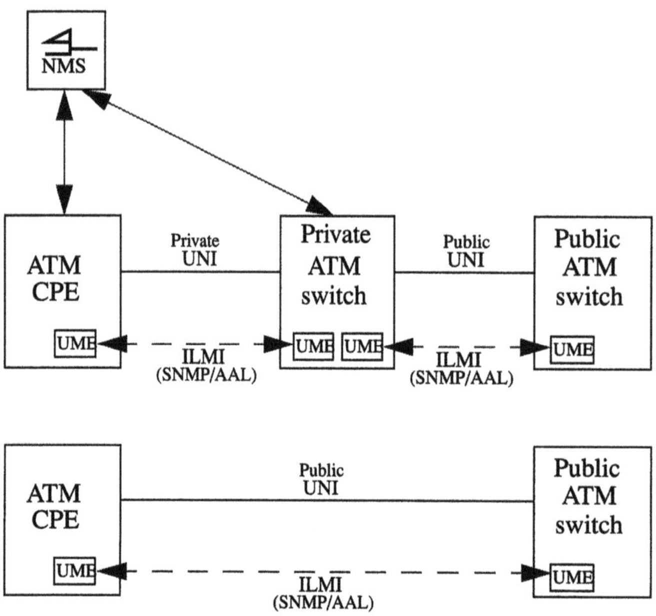

FIGURE 15.2. ILMI in the ATM network.

ILMI Protocol

ILMI supports bidirectional exchange of management information between UNI management entities (UME) which are located both at the ATM end systems and ATM switches. UME terminates ILMI protocol. The ILMI protocol stack is shown in Figure 15.3. A permanently assigned VCC (VPI=0, VCI=16) is used for sending SNMP messages, which are AAL-encapsulated as shown in the ILMI protocol stack. The SNMP messages cannot use more than 1% of the UNI traffic load over ILMI VCC.

ILMI Services and MIB

The ATM UNI management information is represented in an MIB. These MIBs are used for monitoring and controlling of the ATM management information across the UNI. The following types of management information are available in the ATM UNI MIB:

- Physical layer;
- ATM layer;
- ATM layer statistics;
- VP connections;
- VC connections;
- Address registration information.

Physical Layer

The physical layer MIB information includes:

- *Interface index:* This is the index to the physical layer MIB, over which ILMI messages are received;

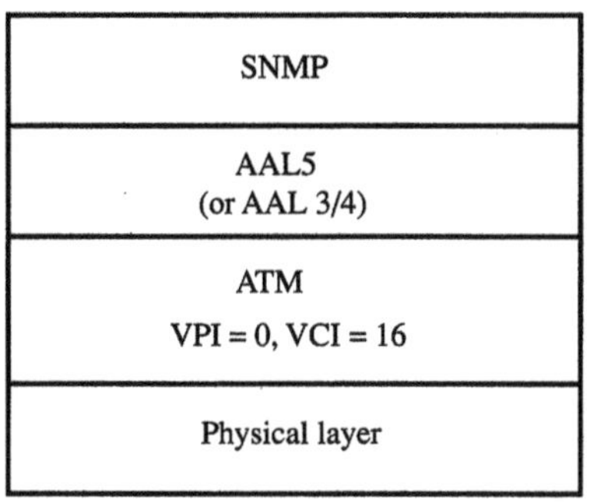

FIGURE 15.3. ILMI protocol stack.

- *Interface address:* 48-bit IEEE 802, or 60-bit ITU-T E.164 (public or private, individual, or group address);
- *Transmission type:* SONET STS-3c, DS-3, 100 Mbps, 155 Mbps;
- *Media type:* Coaxial cable, single-mode fiber, multi-mode fiber, shielded twisted pair, unshielded twisted pair;
- *Operational status:* In-service, out-of-service, loop-back mode.

ATM Layer

The ATM layer MIB information includes:

- *Interface index:* This is the index to the ATM layer MIB, over which ILMI messages are received;
- *VPCs, VCCs:* Maximal number of VPCs and VCCs, which the local interface can support;
- *VPI/VCI address width:* The maximal number of VPI and VCI bits that can be active for this interface;
- *Configured VPCs and VCCs:* Number of configured VPCs and VCCs that the local interface is processing;
- *ATM UNI port type:* Public or private.

ATM Layer Statistics

The ATM layer statistics MIB information includes:

- *Interface index:* This is the index to the ATM layer statistics MIB, over which ILMI messages are received;
- *ATM cells received:* Number of assigned ATM cells received across the ATM UNI. This is a 32-bit wraparound counter;
- *ATM cells dropped on the receive side:* Number of assigned ATM cells dropped across the ATM UNI because of HEC error, unknown VPI/VCI. This is a 32-bit wraparound counter;
- *ATM cells transmitted:* Number of assigned ATM cells transmitted across the ATM UNI. This is a 32-bit wraparound counter.

VP Connections

The VP MIB information includes:

- *Interface index:* This is the index to the VP connections MIB, over which ILMI messages are received;
- *VPI value:* The VPI value for this VPC;
- *Transmit traffic descriptor:* Specification of the conformance definition, source traffic descriptor parameter values that are applicable to the transmit side of this VPC;

- *Receive traffic descriptor:* Specification of the conformance definition, source traffic descriptor parameter values that are applicable to the receive side of this VPC;
- *Operational status:* The status of the VPC, such as end-to-end/local up, down, or unknown;
- *Transmit QoS class:* The QoS defined for the transmit side of this VPC;
- *Receive QoS class:* The QoS defined for the receive side of this VPC.

VC Connections

The VC MIB information includes:

- *Interface index:* This is the index to the VCC MIB, over which ILMI messages are received;
- *VCI value:* The VCI value for this VCC;
- *Transmit traffic descriptor:* Specification of the conformance definition, source traffic descriptor parameter values that are applicable to the transmit side of this VCC;
- *Receive traffic descriptor:* Specification of the conformance definition, source traffic descriptor parameter values that are applicable to the receive side of this VCC;
- *Operational status:* The status of the VCC, such as end-to-end/local up, down, or unknown.
- *Transmit QoS class:* The QoS defined for the transmit side of this VCC;
- *Receive QoS class:* The QoS defined for the receive side of this VCC.

Address Registration

The address registration MIB information includes:

- *Interface index:* This is the index to the address registration information MIB, over which ILMI messages are received.

Internet Protocol

The NMS that deals with the IP uses the specific protocols such as transmission control protocol (TCP), user datagram protocol (UDP), address resolution protocol (ARP) over IP. The ATM UNI and ethernet are used as physical layer for interconnection between NMS and an ATM switch.

Referring to the Figure 15.4, and Figure 15.5, the IP stacks, the file transfer protocol (FTP) application uses TCP. The protocol stack for FTP can be expressed as FTP/TCP/IP/Enet, or FTP/TCP/IP/AAL/ATM. Similarly, the SNMP application uses UDP. The protocol stack for SNMP can be expressed as SNMP/UDP/IP/Enet, or SNMP/UDP/IP/AAL/ATM.

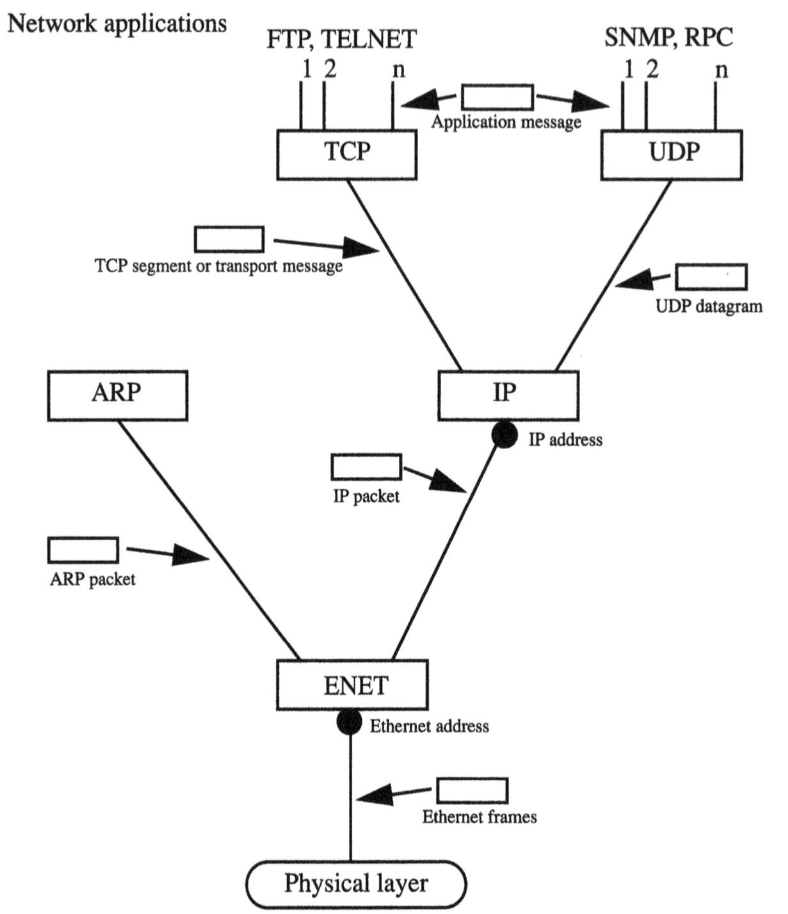

ARP: address resolution protocol
ENET: ethernet
FTP: file transfer protocol
IP: internet protocol
UDP: user datagram protocol
TCP: transmission control protocol
SNMP: simple network management protocol

FIGURE 15.4. IP stack on ethernet.

When either the NMS or the ATM switching equipment needs to communicate with the other, they would normally use SNMP or FTP stacks as described above. In the downward direction, each protocol module adds its header information to create ethernet frames, or ATM cells (all ethernet frames are converted to ATM cells).

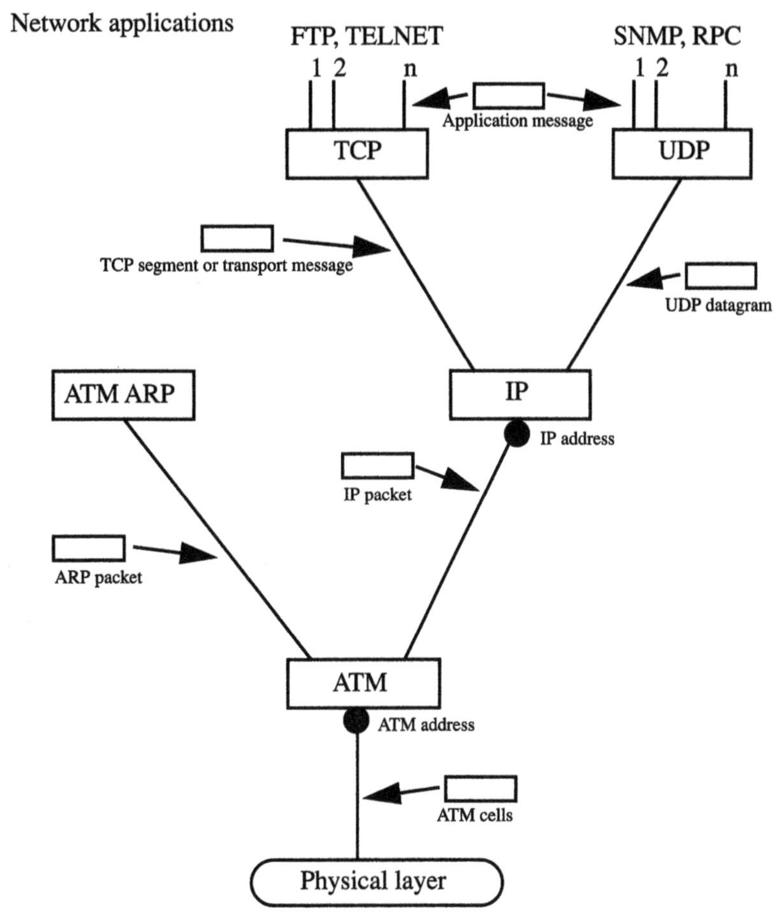

ATM ARP:	ATM address resolution protocol
ENET:	ethernet
FTP:	file transfer protocol
IP:	internet protocol
UDP:	user datagram protocol
TCP:	transmission control protocol
SNMP:	simple network management protocol

FIGURE 15.5. IP stack on ATM UNI.

Then, the ethernet frames or ATM cells travel to the destination ethernet, or ATM port of an ATM switch. These ethernet frames or ATM cells are then transferred to an NMS controller module. The NMS controller either terminates these frames or cells or relays them to the ATM network, so that these cells can travel to the destination NMS controller that is located somewhere in the ATM network.

In the upward direction, ATM cells received from the network or sourced from the NMS controller, that are to be forwarded to the NMS, are converted into ethernet frames (for ethernet interface).

The receiving ethernet driver in the NMS captures the ethernet frames (or ATM cells) and passes them upward to either the ARP (or ATM ARP) module or the IP module. The type field of the ethernet frame determines whether the ethernet frame is passed to the ARP module or to the IP module.

The IP packets are then passed upward to either the TCP module or the UDP module, depending on the protocol filed of the IP header.

The UDP module passes the application message of the UDP datagram upward to the network application module. The port field is analyzed to distinguish among different network applications (0–*n*).

The TCP passes the application message of the TCP segment upward to the network application module based on the value of the port field in the TCP segment header. An example of SNMP over ATM is shown in Figure 15.6. The SNMP server, called an *SNMP agent*, always waits on UDP port 161.

User Datagram Protocol

The UDP offers service to the user's network applications. The network file system (NFS) and SNMP are the examples of network applications. These are connectionless delivery services that do not require guaranteed delivery. UDP preserves message boundary by never joining two messages together, or segmenting a single message.

Transmission Control Protocol

The IP provides connectionless packet delivery service to the transport layer. The error control, flow control, and retransmission (e.g., reliable packet transfer) are not supported by IP. Therefore, the TCP offers service to the user's network applications that are connection oriented and guarantees delivery. FTP, x-window system, and remote copy are examples of network applications. The TCP packetizes the application messages. Some applications may write three times to the TCP port, but the application at the other end may read only one time to retrieve the entire message.

Internet Protocol Address

The IP address is unique for the internet and is four octets long. The IP header contains the IP address, which builds a single logical network from multiple physical networks. The network manager assigns IP addresses to the computers. It depends on the IP network to which the computer is attached. When the computer is relocated to a different network, its IP address is also changed. The network information center (NIC) administers the IP address space.

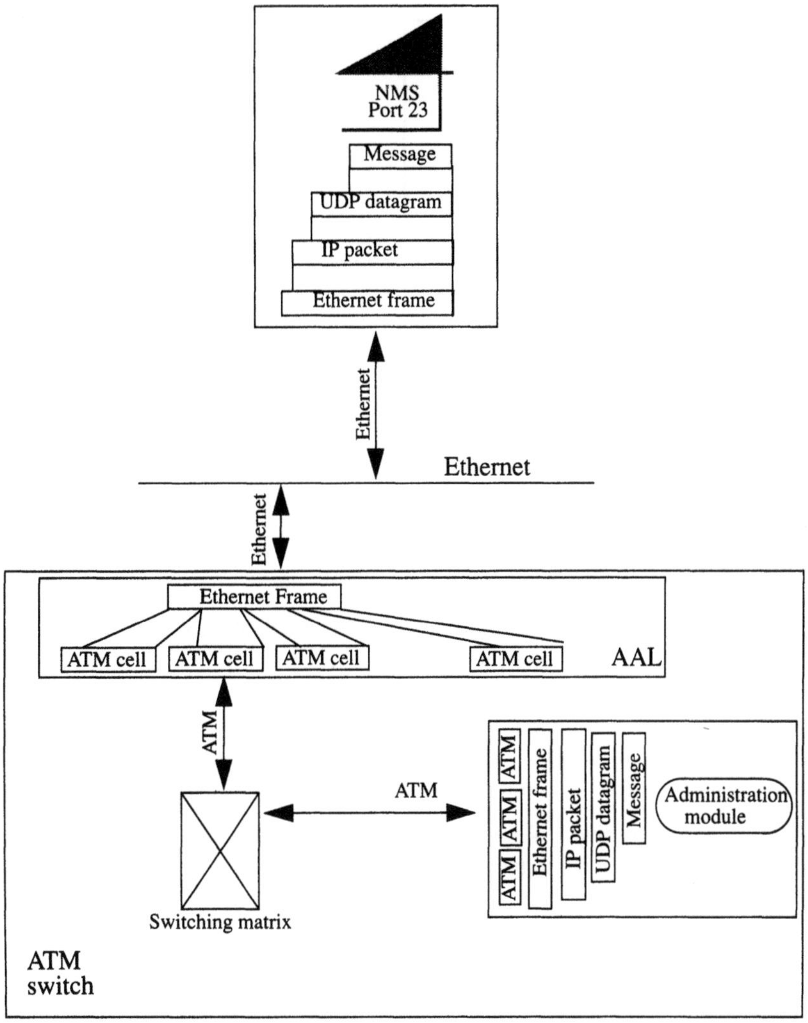

FIGURE 15.6. SNMP over ATM.

Address Resolution Protocol

The outgoing IP packet contains the destination IP address. The ARP is used to translate IP addresses to ethernet addresses. The ARP table contains IP addresses and their corresponding ethernet addresses for all the computers it communicates with. The ARP table is necessary because the IP address and the ethernet address are chosen independently. An example of an ARP table is shown in Table 15.2.

TABLE 15.2. Example of an ARP Table

IP Address	Ethernet Address
123.1.2.1	58-80-28-00-2A-6C
123.1.2.2	58-80-A8-22-2A-BC
123.1.2.3	58-80-4B-45-2A-39
123.1.2.4	58-80-71-21-AA-12

The ARP table is filled automatically by ARP on an as-needed basis. When the translation is missing from the ARP table, the IP packet is stored in the buffer (e.g., queued). The ARP request and response packets update the ARP table, and the queued IP packet is then transmitted.

Ethernet

Carrier sense and multiple access with collision detection (CSMA/CD) is used in ethernet. All devices communicate on a single medium. Only one device can transmit an ethernet frame at any given time. All ethernet devices, however, can receive the ethernet frame simultaneously. If multiple devices try to transmit at the same time, the transmit collision is detected, and these devices must wait a random period before trying to transmit again. See Figure 15.7 for the ethernet structure.

When a frame is received, if the destination address matches with its own ethernet address, then the frame is transferred into its buffer.

The ethernet address is 6 octets long. It is selected by the manufacturer based on the ethernet address space licensed by the manufacturer. When the ethernet hardware interface board changes, the ethernet address changes, too.

Network Management System

The NMS is deployed in the ATM network to monitor, provision, alter, and manage the ATM network resources across several ATM switches. NMS may use traditional IP to communicate with ATM switches. In this example, the communication backbone for NMS is the ethernet. NMS can also use the NNI to communicate with a switch that is not connected directly to it. A group of ATM switches is connected to a single ethernet by the ATM/ethernet converter to communicate with the NMS. The position of the NMS communicating with multiple ATM switches is shown in Figure 15.8.

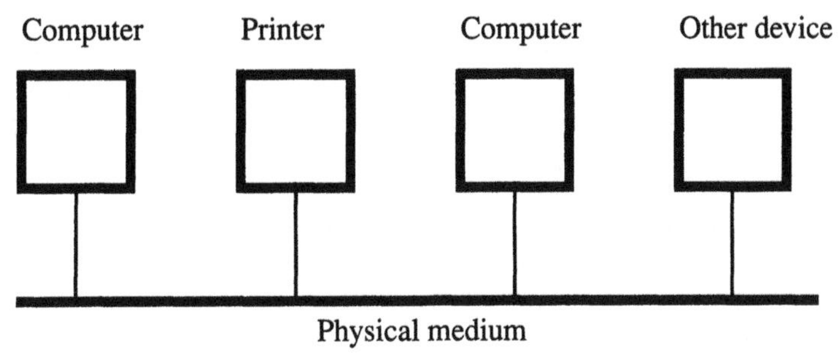

FIGURE 15.7. Ethernet structure.

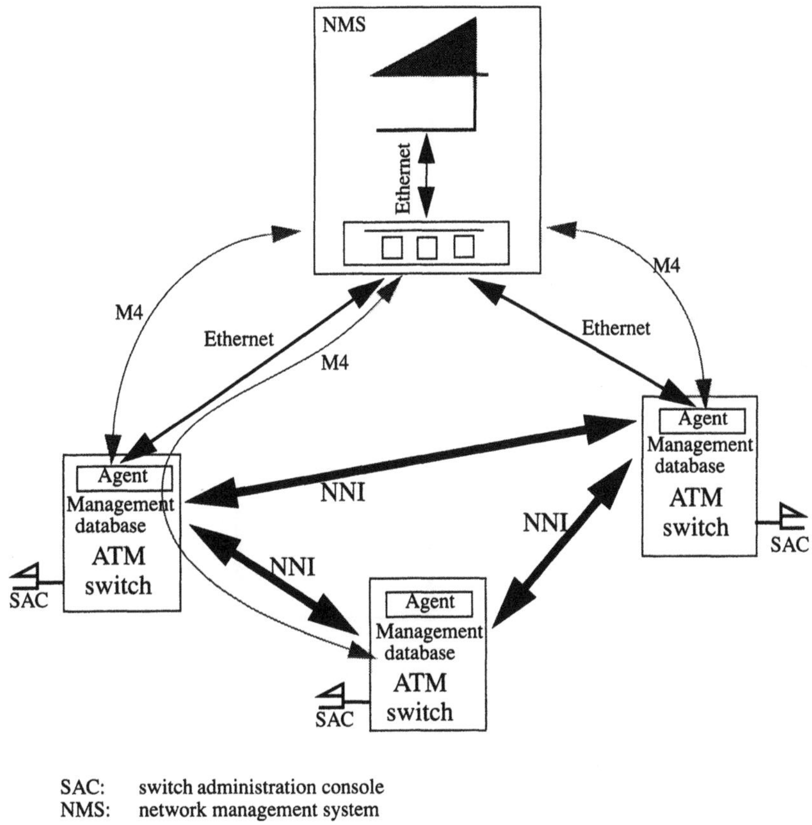

FIGURE 15.8. NMS in the ATM network (direct interface).

Switch Administration Console

The switch administration console (SAC) is used for traffic and congestion control functions, and other configuration and monitoring functions, described previously. It is a local craft terminal. See Figure 15.9 for the SAC in the system. The SAC performs the following functions:

- *Connection management functions*: That establish and configure ATM ports. This is analogous to assigning a telephone number with different features. These functions also test new connections and configurations by sending or receiving simulated traffic (ATM test cells) through these new connections;
- *Fault management functions*: That isolate the faulty equipment in the ATM switch, display them graphically, and reconfigure the connections;
- *Performance management functions*: That run performance tests through connections under test.

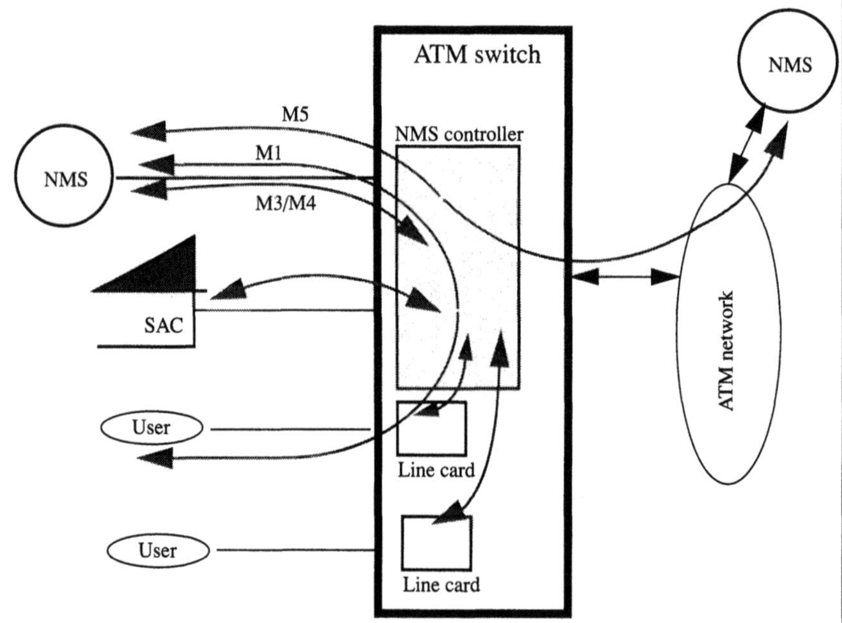

FIGURE 15.9. Switch administration console.

CHAPTER 16

ATM System Blocks

ATM Network Components

The components in an ATM network consist of ATM end stations and a cluster of ATM switches interconnected by the transmission media as shown in Figure 16.1. The main function of the ATM network is to transport ATM cells from one point to the other. The intended information (e.g., user data) that is to be transferred from the originating ATM end station to the destination end station may consist of many thousand octets of data. The ATM cell, however consists of only 53 octets. Therefore, the ATM network requires the end station to packetize the data into 53 octet cells.

The ATM network supports two types of connections:

1. Permanent virtual connections (PVC);
2. Switched virtual connections (SVC).

The PVC is set upon subscription of an ATM service and remains connected permanently. On the other hand, the SVC is set when the ATM end user requests ATM service. The ATM network handles SVC just as dialup connections are handled in today's telephone network. Upon completion of the transmission, the SVC is severed.

In the ATM network, the connection types could be point-to-point or point-to-multipoint. In the point-to-point connection, the originating end station communicates with a single destination end station. In the point-to-multipoint connections, the one end station communicates with multiple end stations. Another type of connection can be established in the ATM network, namely, a broadcast connection. In this type of connection, the originating station broadcasts but does not receive information from any of the destination stations. Multicast connections are also supported in the ATM network. In this type of connection, a single end station can communicate with a group of end stations.

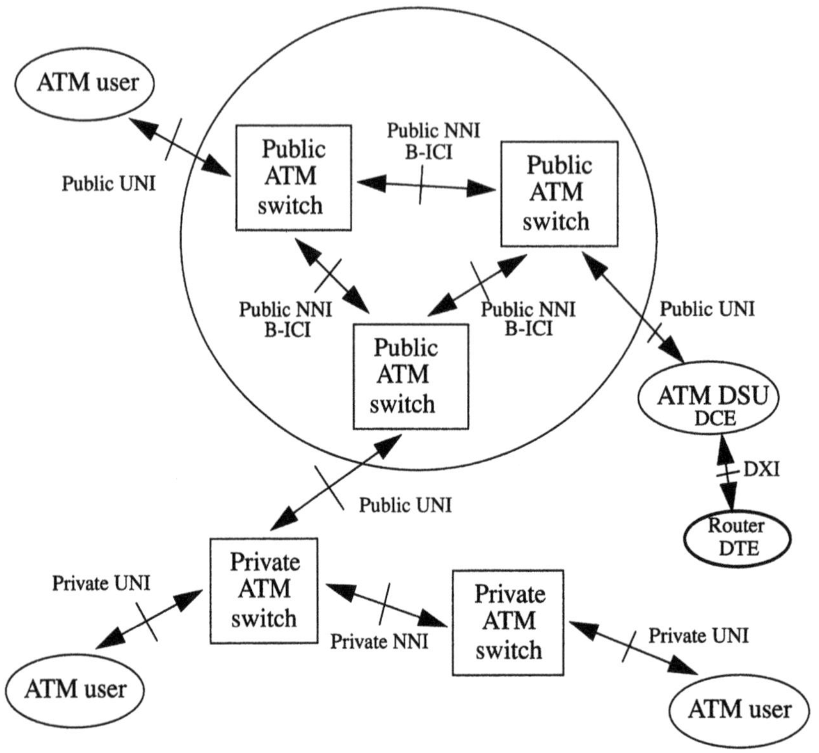

FIGURE 16.1. ATM network.

ATM Network Elements

The ATM network is composed of many components. Components in the ATM network are shown in Figure 16.2.

The following major ATM network equipment is discussed in this chapter:

- ATM end station (e.g., CPE);
- ATM gateway;
- ATM concentrator/cell multiplexer;
- ATM line card;
- Switch administration console;
- ATM administration card (e.g., NMS controller);
- Network management system;
- ATM switching matrix.

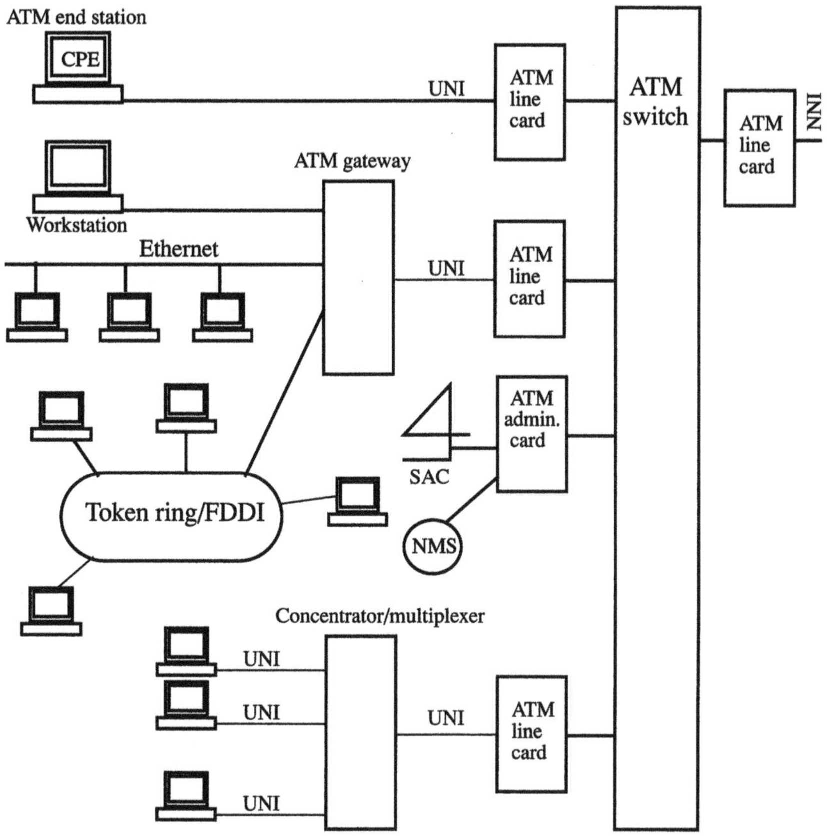

FIGURE 16.2. Components in the ATM network.

ATM End Station

The major function of the ATM end station is to generate ATM cells. The ATM end station has the AAL function built into the system. It converts voice, data, and image information into ATM cells and vice versa. Figure 16.3 shows the block diagram of an ATM end station.

The ATM workstation and the ATM LAN are used as the ATM end station.

ATM Gateway

The major function of the ATM gateway is to convert non-ATM information into ATM cells, which is the only form in which information can be interexchanged

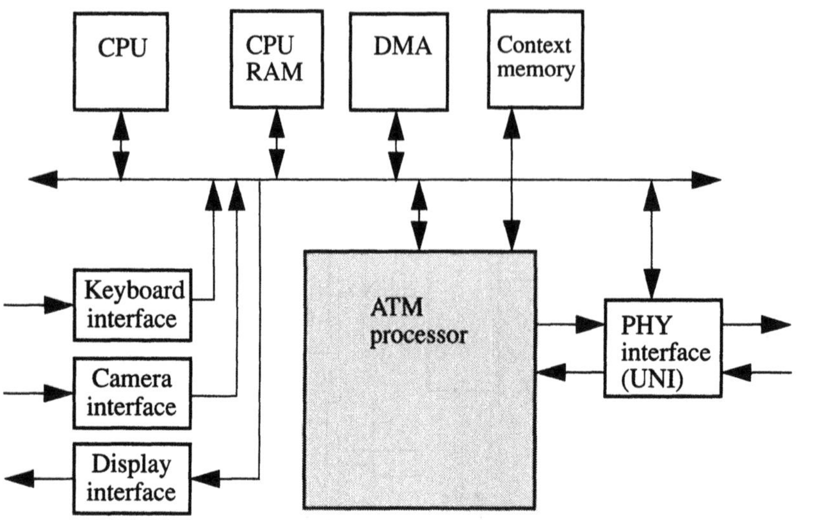

FIGURE 16.3. ATM end station.

through the ATM network. As the gateway to the ATM network, all voice, video, and data services through ATM networks must originate and terminate through an ATM gateway. The commonly used term "ATM gateway" refers to a product that converts non-ATM devices to devices that can interface with the ATM network. The following functions can be performed by an ATM gateway:

- High-speed multiport intelligent bridging function;
- Interconnecting existing LANs to an ATM backbone network;
- Interconnecting the traditional LAN end station to an ATM;
- ATM end station.

The ATM gateway is ideal for a LAN application that requires a higher bandwidth in the backbone network to interconnect different LANs. The ATM gateway provides a smooth migration path to ATM technology without abandoning the existing LAN equipment. Figure 16.4 shows the block diagram of such a system.

ATM gateways can interface with the following devices:

1. LAN:
    - Ethernet;
    - Token ring;
    - FDDI.

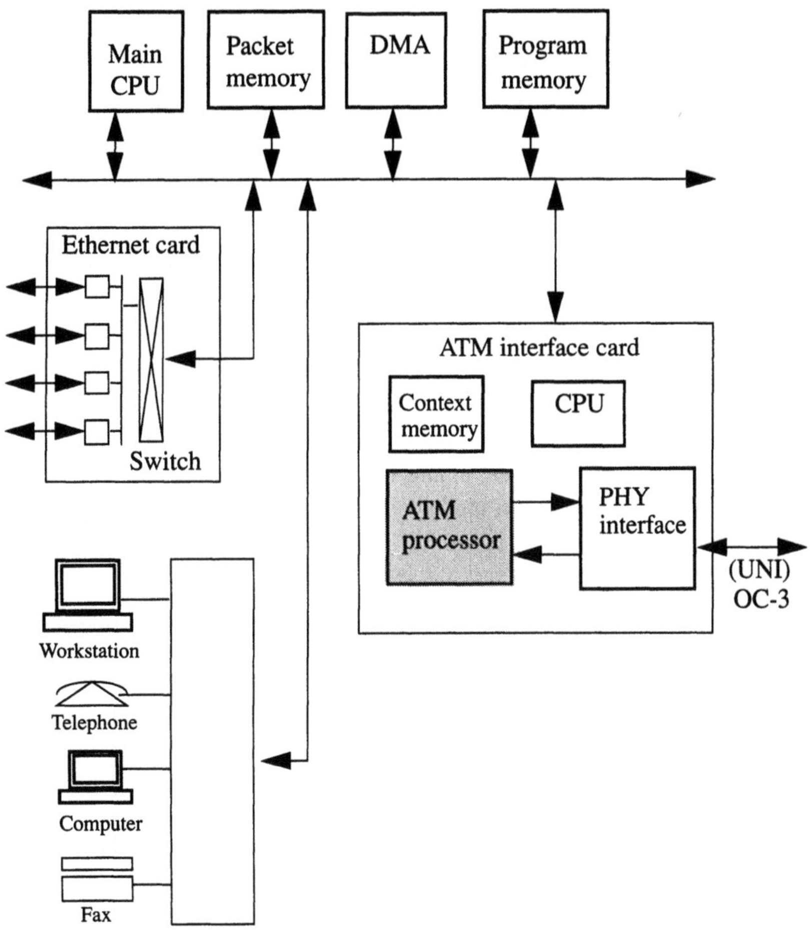

FIGURE 16.4. ATM gateway.

2. Non-ATM devices:
   - Individual workstation;
   - Computer;
   - Telephone;
   - Video terminal;
   - Fax.

The ATM gateway interfaces to the ATM switching network through UNI. The DS-3 and OC-3 physical media are commonly used for this interface.

Concentrator and Cell Multiplexer

The ATM concentrator/cell multiplexer multiplexes several lower bandwidth ATM devices into one high-speed ATM link, and vice versa.

Figure 16.5 shows the block diagram of an ATM concentrator/cell multiplexer.

ATM Line Card

The block diagram of an ATM line card block diagram is shown in Figure 16.6. A more detailed description is given in Chapter 9.

Switch Administration Console

The SAC is a local craft terminal. It interfaces directly to the ATM administration card over a modem or RS232 link. Figure 16.7 highlights the position of the SAC in the system. The functions of this module are discussed in Chapter 15.

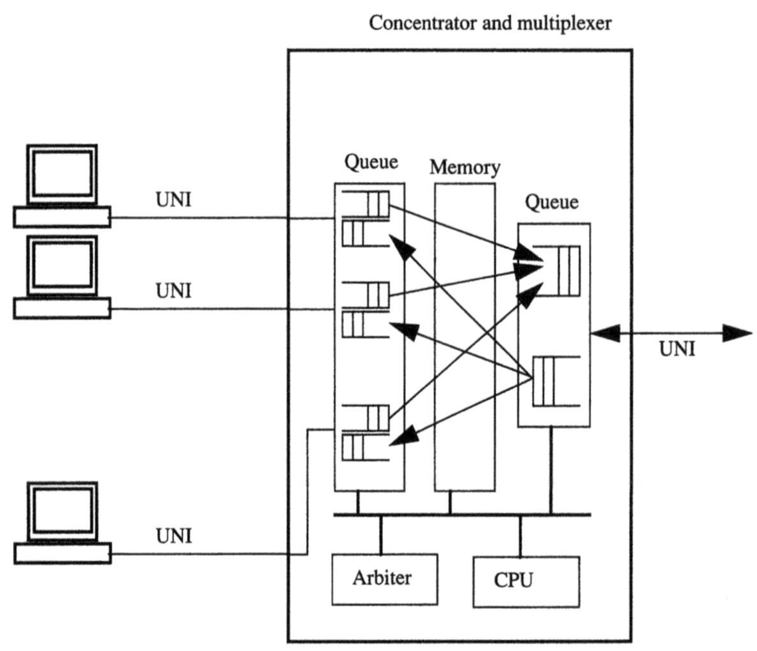

FIGURE 16.5. ATM concentrator and cell multiplexer.

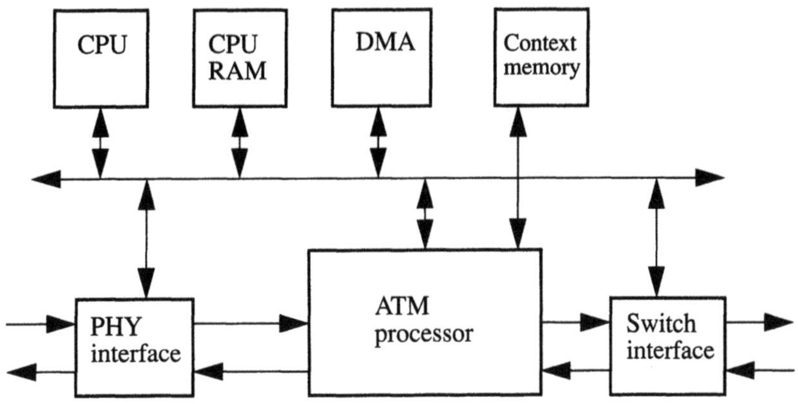

FIGURE 16.6. ATM line card block diagram.

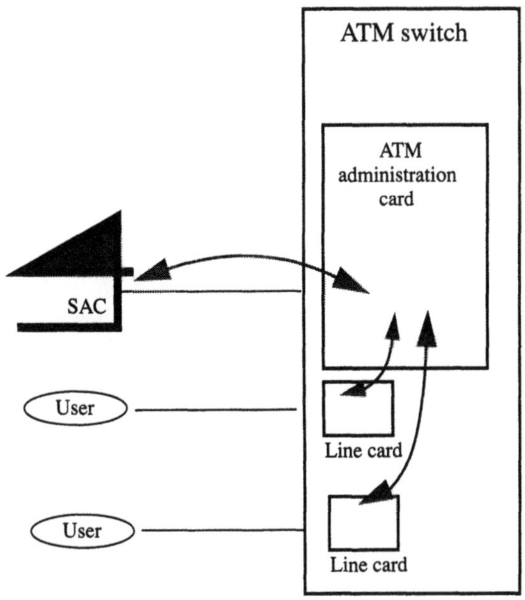

FIGURE 16.7. Switch administration console in the system.

ATM Administration Card

An ATM administration card is used for administration functions. Figure 16.8 highlights the position of the ATM administration card in the system.

This card has the following interfaces:

- NMS interface (ethernet);
- SAC interface (modem or RS232);
- ATM/AAL-5 interface.

Network Management System

The network management system is used for network management functions. Figure 16.9 highlights the position of the network management system in the ATM network. The functions of this module are discussed in Chapter 15.

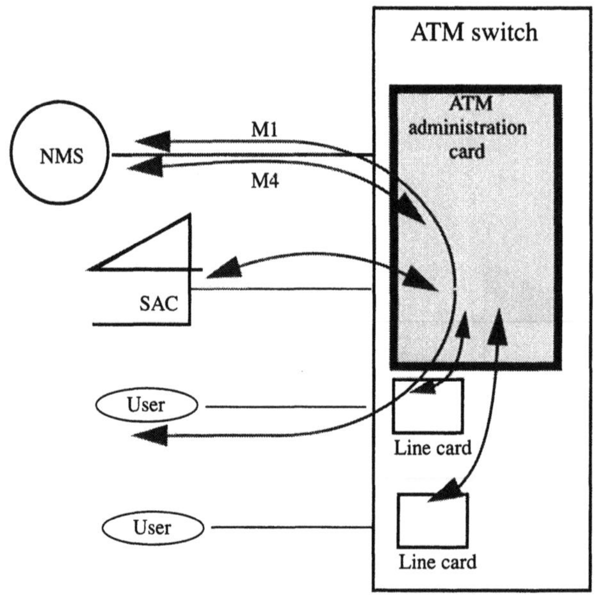

FIGURE 16.8. ATM administration card in the system.

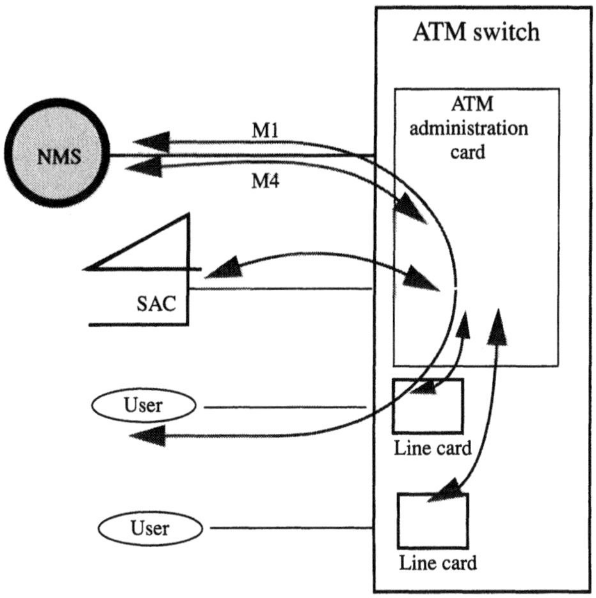

FIGURE 16.9. NMS in the ATM network.

ATM Switching Matrix

The ATM switch consists of multiple physical links and transfers each arriving cell from its ingress link to the egress link according to the routing information contained in the ATM cell header. The switching fabric in the heart of the ATM switch handles the actual routing of the cells. The line card within the ATM switch interfaces between the physical medium and switching fabric. More detailed description is given in Chapter 11.

Bibliography

ATM Forum Documents

www.atmforum.com

B-ICI

ATM Forum, AF-BICI-0013.003, "B-ICI, BISDN Intercarrier Interface," Dec. 1995.

Data Exchange Interface

ATM Forum, AF-DXI-0014.000, "ATM Data Exchange Interface (DXI) Specification," Aug. 1993.

ILMI (Integrated Local Management Interface)

ATM Forum, AF-ILMI-0065.000, "ILMI Specification, Version 4.0," Sep. 1996.

ATM Forum, AF-UNI-0010.002, "ATM User-Network Interface Specification, Version 3.1," 1994.

ATM Forum, "ATM Network Element Management Interface: Functional Requirements and Logical MIB," Version 1.0 (draft), 1994.

ATM Forum, AF-BICI-0015.000, "ATM Physical Medium Dependent Interface Specification for 155 Mb/s over Twisted Pair Cable," Sept. 1994.

ATM Forum, AF-BICI-0016.000, "DS1 Physical Layer Interface Specification," Sept. 1994.

ATM Forum, AF-BICI-0018.000, "Mid-Range Physical Layer Interface Specification for Category 3 Unshielded Twisted Pair," Sept. 1994.

ATM Forum, AF-SAA-0032.000, "Circuit Emulation Service Interoperability Specification," Sept. 1995.

ATM Forum, AF-SAA-0049.000, "Audiovisual Multimedia Services: Video on Demand Specification," Jan. 1996.

ATM Forum, AF-PNNI-0055.000, "Private Network-Network Interface," Sept. 1996.

Signaling

ATM Forum, AF-SIG-0061.000, "ATM User-Network Interface (UNI) Signaling Specification, Version 4.0," July 1996.

ATM Forum, AF-SIG-0076.000, "Signaling ABR Addendum," Jan. 1997.

Traffic Management

ATM Forum, AF-TM-0056.000, "Traffic Management Specification, Version 4.0," Apr. 1996.

ATM Forum, AF-TM-0077.000, "Traffic Management ABR Addendum," Apr. 1996.

Bellcore Documents

www.bellcore.com

Generic Requirements

GR-253-CORE, "Synchronous Optical Network (SONET) Transport Systems: Common Generic Criteria," Bellcore, Dec. 1995.

GR-499-CORE, "Transport Systems Generic Requirements (TSGR): Common Requirements," Bellcore, Dec. 1995.

GR-1110-CORE, "Broadband Switching System (BSS) Generic Requirements," Bellcore, Sept. 1994.

GR-1111-CORE, "Broadband ISDN Access Signaling Generic Requirement," Bellcore, Oct. 1996.

GR-1113-CORE, "Asynchronous Transfer Mode (ATM) & ATM

Adaptation Layer (AAL) Protocol," Bellcore, July 1994.

GR-1114-CORE, "Generic Operations Interface Requirements: ATM Information Model," Bellcore, Sept. 1996.

GR-1115-CORE, "B-ISND Intercarrier Interface (B-ICI) Generic Requirements," Bellcore, Dec. 1995.

GR-1117-CORE, "Generic Requirements for Exchange PVC Cell Relay Service CNM Service," Bellcore, Jan. 1994.

GR-1248-CORE, "Generic Requirements for Operations of ATM Network Elements," Bellcore, Aug. 1996.

GR-1337-CORE, "Multipoint Multimedia Conferencing Control Unit," Dec. 1994.

GR-1837-CORE, "ATM Virtual Path Functionality in SONET Rings–Generic Criteria," Dec. 1994.

GR-2842-CORE, "ATM Service Access Multiplexer Generic Requirements," Bellcore, Feb. 1998.

GR-2845-CORE, "Generic Requirements for the ATM Network and Element Management Layers," Bellcore, June 1994.

GR-2878-CORE, "Generic Requirements for CCS Nodes Supporting ATM High-Speed Signaling Links," Bellcore, Dec. 1997.

GR-2901-CORE, "Video Transport Over Asynchronous Transfer Mode (ATM) Generic Requirements," Bellcore, May 1995.

Technical Advisories

Technical References

ANSI Documents

ANSI T1.110, "Telecommunications—Signaling System No. 7 (SS7)—General Information," 1993.

ANSI T1.102, "Telecommunications—Digital Hierarchy—Electrical Interfaces," 1993.

ANSI T 1.105, "Telecommunications—Digital Hierarchy—Optical Interfaces Rates and Formats Specifications (SONET)," 1991.

ANSI T 1.107a, "Telecommunications—Digital Hierarchy—Formats Specifications (DS3 Format Applications) (supplement to ANSI T1.107-1988)," 1990.

ANSI T1.110, "Telecommunications—Signaling System No. 7 (SS7)—General Information," 1993.

ANSI T1.234, "Telecommunications–Signaling System No. 7 (SS7)–MTP Level 2 and 3 Compatibility Testing," 1993.

ANSI T 1.403, "Telecommunications–Carrier-to-Customer Installation–DS 1 Metallic Interface," 1989.

ANSI T 1.404, "Telecommunications–Network-to-Customer Installation–DS 3 Metallic Interface Specification," 1994.

ANSI T 1.627, "Telecommunications–Broadband ISDN–ATM Layer Functionality and Specification," 1993.

ANSI T 1.629, "Telecommunications–Broadband ISDN–ATM Adaptation Layer 3/4 Common Part Functions and Specification," 1993.

ANSI T 1.630, "Telecommunications–Broadband ISDN–ATM Adaptation Layer for Constant Bit Rate Service Functionality and Specification," 1993.

ANSI T 1.633, "Telecommunications–Frame Relaying Bearer Service Interworking," 1993.

ANSI T 1.634, "Telecommunications–Frame Relaying Service Specific Convergence Sublayer (FR-SSCS)," 1993.

ANSI T 1.635, "Telecommunications–Broadband ISDN–ATM Adaptation Layer 5, Common Part Functions and Specification," 1993.

ANSI T 1.636, "Telecommunications–B-ISDN Signaling ATM Adaptation Layer–Overview Description," 1994.

ANSI T 1.637, "Telecommunications–B-ISDN ATM Adaptation Layer Service Specification Connection-Oriented Protocol (SSCOP)," 1994.

ANSI T 1.638, "Telecommunications–B-ISDN ATM Adaptation Layer–Service Specification Coordination Function for Support of Signaling at the User-to-Network Interface (SSCF at the UNI)," 1994.

ITU Documents

www.itu.com

Series E

ITU-T, recommendation E.164, "The International Public Telecommunication Numbering Plan," May 1997.

Series G

ITU-T, recommendation G.711, "Pulse Code Modulation (PCM) of Voice Frequencies, Blue Book, Fasc. III. 4," Nov. 1988.

ITU-T, recommendation G.723.1,"Dual-Rate Speech Coder for Multimedia Communications Transmitting at 5.3 and 6.3 Kbit/s Test Vectors, Test Sequences and C Reference Code, Annex A," Mar. 1996.

ITU-T, recommendation G.729,"Coding of Speech 8 kbit/s Using Conjugate-Structure Algebraic-Code-Excited Linear-Prediction, Annex A and Annex B," Nov. 1996.

ITU-T, recommendation G.804, "ATM Cell Mapping Into Plesiochronous Digital Hierarchy (PDH)," Nov. 1993.

Series I

ITU-T, recommendation I.121, "Broadband Aspects of ISDN," Apr. 1991.

ITU-T, recommendation I.150, "B-ISDN Asynchronous Transfer Mode Functional Characteristics," Nov. 1995.

ITU-T, recommendation I.211, "B-ISDN Service Aspects (7)," Mar. 1993.

ITU-T, recommendation I.321, "B-ISDN Protocol Reference Model and Its Application," Apr. 1991.

ITU-T, recommendation I.326, "Functional Architecture of Transport Networks Based on ATM," Nov. 1995.

ITU-T, recommendation I.356, "B-ISDN ATM Layer Cell Transfer Performance," Oct. 1996.

ITU-T, recommendation I.361, "B-ISDN ATM Layer Specification," Nov. 1995.

ITU-T, recommendation I.363, "B-ISDN ATM Adaptation Layer Specification," Nov. 1993.

ITU-T, recommendation I.363.1, "B-ISDN ATM Adaptation Layer Specification: Type 1 AAL," Aug. 1996.

ITU-T, recommendation I.363.1, "B-ISDN ATM Adaptation Layer Specification: Type 3/4 AAL," Aug. 1996.

ITU-T, recommendation I.363.1, "B-ISDN ATM Adaptation Layer Specification: Type 5 AAL," Aug. 1996.

ITU-T, recommendation I.365, "B-ISDN ATM Adaptation Layer Sublayer," Nov. 1993.

ITU-T, recommendation I.365.1, "Frame Relaying Service Specific Convergence Sublayer (FR-SSCS)," Nov. 1993.

ITU-T, recommendation I.365.4, "B-ISDN ATM Adaptation Layer Sublayer: Service Specific Convergence Sublayer for HDLC Applications," Aug. 1996.

ITU-T, recommendation I.371, "Traffic Control and Congestion Control in B-ISDN," Aug. 1996.

ITU-T, recommendation I.430, "Basic User-Network Interface—Layer 1 Specification," Nov. 1995.

ITU-T, recommendation I.431, "Primary Rate User-Network Interface—Layer 1 Specification," Mar. 1993.

ITU-T, recommendation I.432, "B-ISDN User-Network Interface—Physical Layer Specification," Aug. 1996.

ITU-T, recommendation I.432.1, "B-ISDN User-Network Interface—Physical Layer Specification: General Characteristics," Aug. 1996.

ITU-T, recommendation I.432.2, "B-ISDN User-Network Interface—Physical Layer Specification: 155.520 kbit/s and 622.080 kbit/s Operation," Aug. 1996.

ITU-T, recommendation I.432.3, "B-ISDN User-Network Interface—Physical Layer Specification: 1544 kbit/s and 2048 kbit/s Operation," Aug. 1996.

ITU-T, recommendation I.432.3, "B-ISDN User-Network Interface—Physical Layer Specification: 51.84 kbit/s Operation," Aug. 1996.

ITU-T, recommendation I.432.3, "B-ISDN User-Network Interface—Physical Layer Specification: 25.600 kbit/s Operation," Aug. 1996.

ITU-T, recommendation I.555, "Frame Relaying Bearer Service Interworking," Mar. 1993.

ITU-T, recommendation I.610: "B-ISDN Operation and Maintenance Principles and Functions," Nov. 1995.

ITU-T, recommendation I.731, "Types and General Characteristics of ATM Equipment," Mar. 1996.

ITU-T, recommendation I.731, "Functional Characteristics of ATM Equipment," Mar. 1996.

ITU-T, recommendation I.751, "Asynchronous Transfer Mode Management of the Network Element View," Mar. 1996.

Series M

ITU-T, recommendation M.20, "Maintenance Philosophy for Telecommunications Network," Oct. 1992.

ITU-T, recommendation M.3010, "Principal for a Telecommunications Management Network," May 1996.

Series Q

ITU-T, recommendation Q.701, "Functional Description of the Message Transfer Part (MTP) of Signaling System No. 7," Mar. 1993.

ITU-T, recommendation Q.710, "Simplified MTP for a Small Systems, Blue Book Fasc. VI.7," Nov. 1988.

ITU-T, recommendation Q.921, "ISDN User-Network Interface Data Link Layer Specification for Basic Call Control," Mar. 1993.

ITU-T, recommendation Q.922, "ISDN Data Link Layer Specification for Frame Mode Bearer Services," Feb. 1992.

ITU-T, recommendation Q.931, "ISDN User-Network Interface Layer 3 Specification for Basic Call Control," Mar. 1993.

ITU-T, recommendation Q.2100, "B-ISDN Signaling ATM Adaptation Layer (SAAL) Overview Description," July 1994.

ITU-T, recommendation Q.2110, "B-ISDN ATM Adaptation Layer Service Specific Connection Oriented Protocol (SSCOP)," July 1994.

ITU-T, recommendation Q.2119, "B-ISDN ATM Adaptation Layer Convergence Function for SSCOP Above the Frame Relay Core Service," July 1996.

ITU-T, recommendation Q.2120, "B-ISDN Meta-Signaling Protocol," Feb. 1995.

ITU-T, recommendation Q.2130, "B-ISDN Signaling ATM Adaptation Layer Service Specific Coordination Function for Support of Signaling at the User-Network Interface (SSCF at UNI)," July 1994.

ITU-T, recommendation Q.2140, "B-ISDN Signaling ATM Adaptation Layer Service Specific Coordination Function for Support for Signaling at the Network Node Interface (SSCF at NNI)," Feb. 1995.

ITU-T, recommendation Q.2144, "B-ISDN Signaling ATM Adaptation Layer (SAAL) Layer Management for the SAAL at the Network Node Interface (NNI)," Oct. 1995.

ITU-T, recommendation Q.2730, "Broadband Integrated Services Digital Network (B-ISDN)—Signaling System No. 7 B-ISDN User Part (B-ISUP)—Supplementary Services," Feb. 1995.

ITU-T, recommendation Q.2761, "Broadband Integrated Services Digital Network (B-ISDN)—Functional Description of the B-ISDN User Part (B-ISUP) of Signaling System No. 7," Feb. 1995.

ITU-T, recommendation Q.2762, "Broadband Integrated Services Digital Network (B-ISDN)—General Functions of Messages and Signals of the B-ISDN User Part (B-ISUP) of Signaling System No. 7," Feb. 1995.

ITU-T, recommendation Q.2763, "Broadband Integrated Services Digital Network (B-ISDN)—Signaling System No. 7 B-ISDN User Part (B-ISUP)—Formats and Codes," Feb. 1995.

ITU-T, recommendation Q.2764, "Broadband Integrated Services Digital Network (B-ISDN)—Signaling System No. 7 B-ISDN User Part (B-ISUP)—Basic Call Procedures," Feb. 1995.

ITU-T, recommendation Q.2931, "Digital Subscriber Signaling No. 2 (DSS 2) User-Network Interface (UNI) Layer 3 Specification for Basic Call/Connection Control," Feb. 1995.

ITU-T, recommendation Q.2961.2, "Support of ATM Transfer Capability in the Broadband Bearer Capability Information Element," June 1997.

ITU-T, recommendation Q.2961.3, "Signaling Capability to Support Traffic Parameters for the Available Bit Rate (ABR) ATM Transfer Capability," Sept. 1997.

ITU-T, recommendation Q.2961.4, "Signaling Capability to Support Traffic Parameters for the ATM Block Transfer (ABT) ATM Transfer Capability," Sept. 1997.

Series X

ITU-T, recommendation X.213, "Information Technology—Open Systems Interconnection—Network Service Definition," Nov. 1995.

IETF

RFC 1157, "Simple Network Management Protocol (SNMP)," May 1990.

RFC 1483, "Multiprotocol Encapsulation Over ATM Adaptation Layer 5," July 1993.

RFC 1490, "Multiprotocol Interconnect Over Frame Relay," July 1993.

RFC 1577, "Classical IP and ARP Over ATM," Jan. 1994.

Other Documents

Van de Goor, A. J., *Computer Architecture and Design*, Addison-Wesley Publishing Company, 1989.

McDysan, D. E., and D. L. Spohn, *ATM Theory and Application*, McGraw-Hill, 1994.

Fredkin, E., "Trie Memory," *Communication of the ACM*, vol. 3, no. 9, pp. 490–499, Sept. 1960.

Maly, K.,"Compressed Tries," *Communication of the ACM*, vol. 19, no. 7, pp. 409–415, July 1976.

Rahman, M., "An Architecture for Locating Mobile Units of a Cellular Mobile System Through the Intelligent Network (IN)," Ph.D. dissertation, Dept. of Electrical Engineering, Southern Methodist University, 1994.

Onvural, R. O., *Asynchronous Transfer Mode Networks—Performance Issues*, Artech House, 1994.

Handel, R., M. N. Huber, and S. Schroder, *ATM Networks, Concepts, Protocols, Applications*, Addison-Wesley, 1994.

Tong-Bi-Pei, and C. Zukowski, "Putting Routing Tables in Silicon," *IEEE Network Magazine*, Jan. 1992.

Socolofsky, T., and C. Kale, "A TCP/IP Tutorial," Network Working Group Request for Comments, 1180, Jan. 1991.

About the Author

Dr. Mohammad A. Rahman is a senior systems engineer with DSC Communications Corporation and has been active in the telecommunications field for more than 14 years. He has also been an adjunct professor in the Electrical Engineering Department, Southern Methodist University, since 1995. He is teaching a graduate course called "ATM Systems in Telecommunications."

Index

B

C

X

The Artech House Telecommunications Library

Vinton G. Cerf, Series Editor

Access Networks: Technology and V5 Interfacing, Alex Gillespie

Advanced High-Frequency Radio Communications, Eric E. Johnson, Robert I. Desourdis, Jr., et al.

Advanced Technology for Road Transport: IVHS and ATT, Ian Catling, editor

Advances in Computer Systems Security, Vol. 3, Rein Turn, editor

Advances in Telecommunications Networks, William S. Lee and Derrick C. Brown

Advances in Transport Network Technologies: Photonics Networks, ATM, and SDH, Ken-ichi Sato

An Introduction to International Telecommunications Law, Charles H. Kennedy and M. Veronica Pastor

Asynchronous Transfer Mode Networks: Performance Issues, Second Edition, Raif O. Onvural

ATM Switches, Edwin R. Coover

ATM Switching Systems, Thomas M. Chen and Stephen S. Liu

Broadband: Business Services, Technologies, and Strategic Impact, David Wright

Broadband Network Analysis and Design, Daniel Minoli

Broadband Telecommunications Technology, Byeong Lee, Minho Kang and Jonghee Lee

Cellular Mobile Systems Engineering, Saleh Faruque

Cellular Radio: Analog and Digital Systems, Asha Mehrotra

Cellular Radio: Performance Engineering, Asha Mehrotra

Cellular Radio Systems, D. M. Balston and R. C. V. Macario, editors

CDMA for Wireless Personal Communications, Ramjee Prasad

Client/Server Computing: Architecture, Applications, and Distributed Systems Management, Bruce Elbert and Bobby Martyna

Communication and Computing for Distributed Multimedia Systems, Guojun Lu

Communications Technology Guide for Business, Richard Downey, et al.

Community Networks: Lessons from Blacksburg, Virginia, Andrew Cohill and Andrea Kavanaugh, editors

Computer Networks: Architecture, Protocols, and Software, John Y. Hsu

Computer Mediated Communications: Multimedia Applications, Rob Walters

Computer Telephone Integration, Rob Walters

Convolutional Coding: Fundamentals and Applications, Charles Lee

Corporate Networks: The Strategic Use of Telecommunications, Thomas Valovic

The Definitive Guide to Business Resumption Planning, Leo A. Wrobel

Digital Beamforming in Wireless Communications, John Litva and Titus Kwok-Yeung Lo

Digital Cellular Radio, George Calhoun

Digital Hardware Testing: Transistor-Level Fault Modeling and Testing, Rochit Rajsuman, editor

Digital Switching Control Architectures, Giuseppe Fantauzzi

Digital Video Communications, Martyn J. Riley and Iain E. G. Richardson

Distributed Multimedia Through Broadband Communications Services, Daniel Minoli and Robert Keinath

Distance Learning Technology and Applications, Daniel Minoli

EDI Security, Control, and Audit, Albert J. Marcella and Sally Chen

Electronic Mail, Jacob Palme

Enterprise Networking: Fractional T1 to SONET, Frame Relay to BISDN, Daniel Minoli

Expert Systems Applications in Integrated Network Management, E. C. Ericson, L. T. Ericson, and D. Minoli, editors

FAX: Digital Facsimile Technology and Applications, Second Edition, Dennis Bodson, Kenneth McConnell, and Richard Schaphorst

FDDI and FDDI-II: Architecture, Protocols, and Performance, Bernhard Albert and Anura P. Jayasumana

Fiber Network Service Survivability, Tsong-Ho Wu

Future Codes: Essays in Advanced Computer Technology and the Law, Curtis E. A. Karnow

Guide to ATM Systems and Technology, Mohammad A. Rahman

Guide to Telecommunications Transmission Systems, Anton A. Huurdeman

A Guide to the TCP/IP Protocol Suite, Floyd Wilder

Implementing EDI, Mike Hendry

Implementing X.400 and X.500: The PP and QUIPU Systems, Steve Kille

Inbound Call Centers: Design, Implementation, and Management, Robert A. Gable

Information Superhighways Revisited: The Economics of Multimedia, Bruce Egan

Integrated Broadband Networks, Amit Bhargava

International Telecommunications Management, Bruce R. Elbert

International Telecommunication Standards Organizations, Andrew Macpherson

Internet E-mail: Protocols, Standards, and Implementation, Lawrence Hughes

Internetworking LANs: Operation, Design, and Management, Robert Davidson and Nathan Muller

Introduction to Document Image Processing Techniques, Ronald G. Matteson

Introduction to Error-Correcting Codes, Michael Purser

An Introduction to GSM, Siegmund Redl, Matthias K. Weber and Malcom W. Oliphant

Introduction to Radio Propagation for Fixed and Mobile Communications, John Doble

Introduction to Satellite Communication, Bruce R. Elbert

Introduction to T1/T3 Networking, Regis J. (Bud) Bates

Introduction to Telephones and Telephone Systems, Second Edition, A. Michael Noll

Introduction to X.400, Cemil Betanov

LAN, ATM, and LAN Emulation Technologies, Daniel Minoli and Anthony Alles

Land-Mobile Radio System Engineering, Garry C. Hess

LAN/WAN Optimization Techniques, Harrell Van Norman

LANs to WANs: Network Management in the 1990s, Nathan J. Muller and Robert P. Davidson

Minimum Risk Strategy for Acquiring Communications Equipment and Services, Nathan J. Muller

Mobile Antenna Systems Handbook, Kyohei Fujimoto and J. R. James, editors

Mobile Communications in the U.S. and Europe: Regulation, Technology, and Markets, Michael Paetsch

Mobile Data Communications Systems, Peter Wong and David Britland

Mobile Information Systems, John Walker

Multimedia Communications Networks: Technologies and Services, Mallikarjun Tatipamula and Bhumip Khasnabish

Networking Strategies for Information Technology, Bruce Elbert

Packet Switching Evolution from Narrowband to Broadband ISDN, M. Smouts

Packet Video: Modeling and Signal Processing, Naohisa Ohta

Performance Evaluation of Communication Networks, Gary N. Higginbottom

Personal Communication Networks: Practical Implementation, Alan Hadden

Personal Communication Systems and Technologies, John Gardiner and Barry West, editors

Practical Computer Network Security, Mike Hendry

Principles of Secure Communication Systems, Second Edition, Don J. Torrieri

Principles of Signaling for Cell Relay and Frame Relay, Daniel Minoli and George Dobrowski

Principles of Signals and Systems: Deterministic Signals, B. Picinbono

Private Telecommunication Networks, Bruce Elbert

Radio-Relay Systems, Anton A. Huurdeman

RF and Microwave Circuit Design for Wireless Communications, Lawrence E. Larson

The Satellite Communication Applications Handbook, Bruce R. Elbert

Secure Data Networking, Michael Purser

Service Management in Computing and Telecommunications, Richard Hallows

Signaling in ATM Networks, Raif O. Onvural, Rao Cherukuri

Smart Cards, José Manuel Otón and José Luis Zoreda

Smart Card Security and Applications, Mike Hendry

Smart Highways, Smart Cars, Richard Whelan

Successful Business Strategies Using Telecommunications Services, Martin F. Bartholomew

Super-High-Definition Images: Beyond HDTV, Naohisa Ohta, et al.

Telecommunications Deregulation, James Shaw

Television Technology: Fundamentals and Future Prospects, A. Michael Noll

Telecommunications Technology Handbook, Daniel Minoli

Telecommuting, Osman Eldib and Daniel Minoli

Telemetry Systems Design, Frank Carden

Teletraffic Technologies in ATM Networks, Hiroshi Saito

Toll-Free Services: A Complete Guide to Design, Implementation, and Management, Robert A. Gable

Transmission Networking: SONET and the SDH, Mike Sexton and Andy Reid

Troposcatter Radio Links, G. Roda

Understanding Emerging Network Services, Pricing, and Regulation, Leo A. Wrobel and Eddie M. Pope

Understanding GPS: Principles and Applications, Elliot D. Kaplan, editor

Understanding Networking Technology: Concepts, Terms and Trends, Mark Norris

UNIX Internetworking, Second Edition, Uday O. Pabrai

Videoconferencing and Videotelephony: Technology and Standards, Richard Schaphorst

Voice Recognition, Richard L. Klevans and Robert D. Rodman

Wireless Access and the Local Telephone Network, George Calhoun

Wireless Communications in Developing Countries: Cellular and Satellite Systems, Rachael E. Schwartz

Wireless Communications for Intelligent Transportation Systems, Scott D. Elliot and Daniel J. Dailey

Wireless Data Networking, Nathan J. Muller

Wireless LAN Systems, A. Santamaría and F. J. López-Hernández

Wireless: The Revolution in Personal Telecommunications, Ira Brodsky

Writing Disaster Recovery Plans for Telecommunications Networks and LANs, Leo A. Wrobel

X Window System User's Guide, Uday O. Pabrai

www.ingramcontent.com/pod-product-compliance
Ingram Content Group UK Ltd.
Pitfield, Milton Keynes, MK11 3LW, UK
UKHW051457290926
15471UKWH00008B/1251